MYSTERIES

UNEXPLAINED

PROF. R V M CHOKKALINGAM

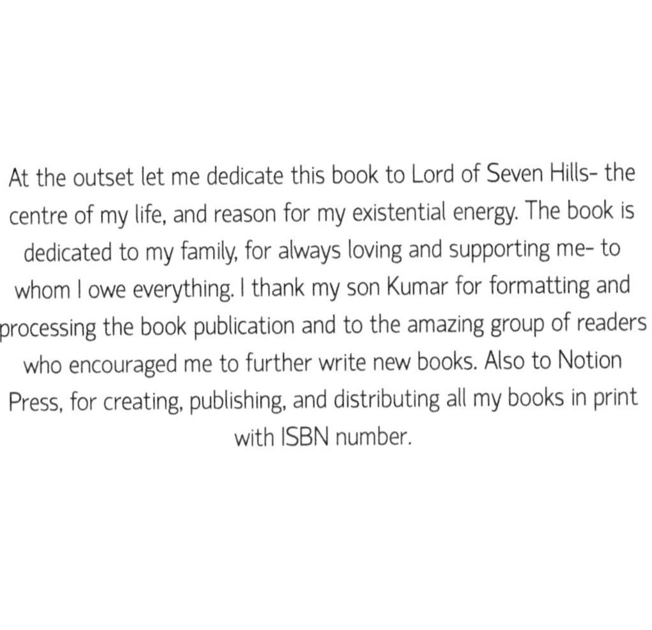

At the outset let me dedicate this book to Lord of Seven Hills- the centre of my life, and reason for my existential energy. The book is dedicated to my family, for always loving and supporting me- to whom I owe everything. I thank my son Kumar for formatting and processing the book publication and to the amazing group of readers who encouraged me to further write new books. Also to Notion Press, for creating, publishing, and distributing all my books in print with ISBN number.

Contents

CHAPTER ONE

Why Mysteries Matter?

Mystery, in the simplest sense, is something we do not know the answer to. In other words, mystery is about trying to find answer. Mystery means the acceptance of things we do not know or understand. Mystery is awareness that there is something more to life. The more we know, the more mystery there is. Mysteries often engage us with emotion, and usually via vicarious excitement and fear. Mysteries help us expand our horizons, and give us a greater repertoire for interacting with the world. Mysteries trigger curiosity, along with surprise, interest, doubt, and other effective states. Mysteries offer so much richness, and enjoy the sense of supernatural. Mystery provides a means to explore and possibly understand a complex world often in the darkest terms. The most beautiful experience we can have is the mysterious. The sensation can be uncanny and a bit thrilling. When a mystery presents itself, we have a felt sense of its presence. Mystery enlarges our awareness that the unexplainable exists as a governing force that works for harmony in the universe.

We live in a world where we try to come up with an explanation for almost everything. Therefore, mysteries that defy our basic reasoning can quickly catch our attention, arouse our curiosity, and stimulate our imagination. We love mysteries because there is an uncontrollable thrill while we encounter something strange and unexpected. Some of the world's historical enigma gives us a chance of conjuring up things that are either unrealistic or surprise. Mysteries matter, perhaps more than ever before. Mystery intersects with our lives in dreams and transports us to another existence that is equally real. In dreams, the profound mysteries of symbol, image, and meaning combine. Mysteries are about trying to find answers and help us expand our horizon. Religious myths attempt to explain the unknown with the unknowable, while science attempts to explain the unknown with the knowledge. Though we strive toward more knowledge, we must understand that we are, and will remain, surrounded by mystery. The urge to go beyond the boundaries of the known that feeds our creative impulse, is often flirting with mystery.

Science has come a long way, yet we still do not understand everything about our universe- or ourselves. The book takes the reader on an incredible rollercoaster exploring the top unsolved mysteries in science that have kept the world constantly curious. Though scientists may have theories about the natural phenomena, no one can say for certain about: Which drives the evolution? What is the real nature of time? Where did mathematics come from? Are we alone in the universe? How did the human language evolve? And other famous unsolved mysteries. These are the most important, interesting, and awe-inspiring unanswered questions in science that the book does narrate. These are the top scientific mysteries for the 21ˢᵗ century scientists to solve. There are also picks in the book about the most profound open questions on Dark Energy, Dark Matter, Black Hole, Consciousness, Life definition etc. Some of these mysteries are fundamental and commonplace that it is surprising that we do not know more. Despite new discoveries every day, the human species still live in a world of various elements in secrecy. For centuries, scientists have been struggling to answer seemingly simple questions like this.

Evolution is a scientific theory supported by an overwhelming amount of evidence. One of the greatest mysteries in biology at the moment is whether natural selection is the only process capable of generating organism complexity and whether there are other properties of matter that also come into play. It seems highly paradoxical to assert that time is unreal, and that all statements which involve its reality are erroneous. Yet in the all ages the belief in the unreality of time has proved singularly attractive. There is uncanny accuracy in the way mathematics can reveal the secrets of the universe, making it seem inherent part of the nature. We are not sure whether mathematics is humankind's clever trick, or deeply embedded language of cosmos. The question Are we alone? has plagued humankind since time immemorial. Most people believe that outer space aliens exist, and they have visited our planet. Language is a reflection of us, and it gives voice to our thoughts. But the origin of language is an intriguing question, to which we may never have a complete answer. Thought without symbols, and life without language are virtually impossible for most modern humans to fathom.

The world is full of mysteries some of which have left humans confused for centuries. There is nothing like a good mystery; especially when it has been unsolved for a very long period of time. Our planet earth is a wondrous place of mysteries with mythical origins, or unexplained phenomena. The planet is dotted with places that are surreal, eerie, and sometimes downright bizarre. Many of these mysteries may have different explanations, but they largely remain issues that humans have continued to decipher. These fascinating places still remain as puzzles to mankind. There are some very ancient and weird places around the world ranging from enigmatic stone statues of Easter Island, to unbelievable archaeological findings, to landing sites of visitors from other world. No matter if we are conspiracy theorists, some places are perfect to indulge in those weird and wonderful interests. Some of the spookiest locations are jaw-dropping as they have been never resolved. Many of these places are natural spectacles, which are truly mysterious in themselves. The intricate designs, and natural patterns, and peculiar contours of the corners of the world make us to seek explanations.

The Great Pyramid of Giza has been captivating mankind for thousands of years. The only one of the seven wonders of the ancient world-that is still intact. Area 51 is the Air Force facility located within Nevada Test and Training Range has captured the imagination of conspiracy theorists. The Polish Crooked Forest with hundreds of peculiar pine trees grew with an almost 90-degree bend at their bases, making them look like fishing hooks. The Devil's Bridge in Germany is one of the most stunning parabolic bridges in the world that forms a perfect circle with its own reflection in the water below, a feat only deemed possible with some other worldly assistance. Magnetic Hills of Leh in India are optical illusions in which a road that looks like it's sloping uphill due to the surrounding landscape is actually sloping downhill, so cars, buses, and other vehicles appear to roll uphill in defiance of gravity. Marfa Lights in Texas appear as bright, colorful orbs pulsating or darting through the dark and wide Texas skies. What the lights are and how they are created remains a mystery. Researchers have tried to crack their secrets but cannot seem to agree on the answers.

For instance, the UFO sightings carved on the walls of ancient Egyptian Pyramids seem to suggest something extraordinarily suspicious: did they make contact with aliens? Did aliens help them build the Pyramids? What if they were actually descendents of aliens? Sacred mysteries are the areas of super-natural phenomena associated with a divinity. The secret things of God are the mysteries that God knows everything which the humanity is desperately trying to know. We have coped with encounters with the mysteries in our life. Albert Einstein once wrote: "the most beautiful experience we can have is the mysterious". In science, many unsolved mysteries may contain important insights into a specific area of study. Such mysteries require critical insights, and creative mind. The process of solving scientific mysteries can oftentimes lead to many unexpected results and insights. As our knowledge grows, the mystery we face seems to expand. Perhaps the urge to solve an intriguing mystery is just part of human nature, encoded in our DNA to make sense of the world. Cosmic quandaries and mysterious secrets about space that remains as yet unsolved.

Our ancestors built many amazing cities and temples, and carved so much fascinating artwork, but often the methods and reasons behind what they did have been lost in the mists of time. A far greater treasure lies hidden beneath the surface of the ancient mysteries. Some of the mysteries are natural occurrences, but some are beyond explanation. There is always fear of the unknown where there is mystery. Everyone wants to know what is in the unknown. Where any answer is possible for a mystery, all answers may become meaningless. The answers we seek in mysteries are not always the answers we want. Also the earth is full of weird places with some of the most alien landscape. One common conspiracy theory about the Pyramids is that they and other ancient structures were built by aliens, despite much evidence to the contrary. Aside from depiction of UFOs in media, UFOs have also become part of our folk culture, and a part of the mythology. It is essential to entertain and explore new ideas, however strange, while at the same time testing and evaluating the validity of those ideas. It can be hard to believe, but there are still many world mysteries for which there are no obvious or easy answers.

We do not seem to know the mystery – where the laws of nature come from? Even physicists are troubled by the question of how something abstract and outside of the universe can govern the motions of bodies inside it. Another interesting thing is, of course, that there may be many different theories describing the same thing. We do not know that the equations in the laws of nature with phrased mathematics give reliable predictions are prescriptions, or descriptions, or patterns. Natural laws are curious things and their understanding is that supernatural being brings about the laws. The theological notion is that God made the laws that govern the universe, and physicists involve n the task of discovering those laws. Physicists come up with equations that express the laws of nature that describe phenomena occurring in nature and have been proven by the scientific method. Scientific laws are, in a sense, the rules of the game that nature plays. Laws of nature are determined by fundamental forces within nature, and are concise description of natural phenomena. A law of nature can be a simple statement in words as well as a mathematical equation.

The mystery is that the laws of nature express something that seems clearly true. They are independent of us and appear to work over and above all other objects to which they apply. While physicists come up with equations that express the laws of nature, philosophers think about what a law of nature actually is. The laws of nature, in some sense, are not part of the universe, but they actually govern, or dictate how nature should behave. The laws of nature are fundamental elements of our reality, separate from all the material stuff out there, yet they dictate natural processes. The most incomprehensible thing about the universe is that it is comprehensible. The reality is simply a distribution of stuff endowed with certain properties through time and space. Scientists look at all the information they have about the universe, and try to come up with a system, a theory to describe that information. If the laws of physics are to be real laws, they have to be good anywhere and anytime in the universe. The mystery of laws of nature is whether they are descriptive, or prescriptive? The one thing a scientific law does not explain is why the phenomenon exists, or what causes it?

Mystery is simply a hidden reality or secret. Mystery signifies the value of knowledge that surpasses human reason and hovers somewhere beyond the limits of current human knowledge. Human beings are remarkably prone to supernatural beliefs and in particular to beliefs in invisible agents. It is a real mystery why are we drawn to such beliefs. Supernatural beliefs are hardwired into our brain from birth. But almost everyone entertains some form of irrational beliefs even if they are not religious. A majority of us believe in supernatural- not just because we want to, but because we have to. If there are not really good grounds for believing such beings exist, however, why people believe them. When people are asked to justify their beliefs in such invisible beings, they often appeal to two things such as the reports of sightings and miraculous events. Invisible agents provide quick, convenient explanations for events that might otherwise strike us as deeply mysterious. A belief in the supernatural can give a deep sense of connection with the past and with each other. Naturally knowable truths remain obscure, and mysterious power cannot be construed by any of the known theories.

The essence of empirical science is that nature always has the last word. Machines developed by technology determine what we can measure and learn about the universe. It then follows that we only have limited access to nature through our machines and tools, and more subtly through our restricted methods of investigation. Therefore our knowledge of natural world is necessarily limited. If large portions of the world remain unseen or inaccessible to us, we can ever hope to grasp it in its totality. So the version of reality we might call true at one time will not remain true at another- the ultimate truth is elusive. To go beyond the known, science has to take intellectual risks, making assumptions based on intuition and personal prejudice. What we see of the world is only a silver of what is out there, but there is much that is invisible to the eye, because much of nature remains hidden from us. Our view of the world is based only on the fraction of reality that we can measure and analyze. Science, as our narrative describing what we see and what we conjure exists in the natural world. It is thus necessarily limited, telling only part of the story. The map of what we call reality is an ever-shifting mosaic of ideas. Each new discovery seems to unlock a Pandora box of even bigger and deeper mysteries.

• 9 •

CHAPTER TWO

Which Drives Evolution?

Creationism is still very much a living phenomenon today and the same is true of its offspring – Intelligent Design. Modern Creationist thought is obviously generated by fundamental theology. Indeed, theologians have reconciled human evolution with belief in a transcendent and immanent Creator God. Creationists think that a full understanding of organic world demands the invocation of some force beyond nature. In its basic generic sense, Creationism refers to any view that rejects evolution and favors supernatural Creator. Creationists suggest that every tree that does not bring forth good fruit is hewn down and cast into fire could be taken as an example of the process of natural selection. One of the most consistent arguments by the new Creationists is that the theory of evolution amounts to naturalist or materialist dogma. Creationists say that the scientific method fails to yield an accurate representation of the world. Creationists argue that the diversity of life on earth is best accounted for by the action of an Intelligent Force. But scientists feel that the Creationist's failure as a scientific endeavor and their misconception about the scientific practice are the main concern.

Creationism is the belief that the universe and living organisms originate from specific of divine creation. Creationism is a religiously motivated worldview in denial of biological evolution that has been very resistant to change. Religious belief adds another dimension to the way people interact with evolutionary science. Creationism is the rejection of evolution in favor of supernatural design that comes in many varieties. Creationism in its many forms insists that everything in nature was created by a deity, which is slowly extending to reach. A creationist is someone who believes in God-the absolute creator of heaven and earth out of nothing by an act of free will. Creationists clearly operate from a teleological worldview and often claim that the evolutionary theory eliminates purpose in human existence. Creationists often justify their rejection of biological evolution by claiming that the methodologies and interpretations of evolutionary scientists are flawed. Creationists point to the anthropic principle- the idea that fundamental physical constants of the universe must be as they are for life to exist, as evidence that the organisms of the universe requires an intelligent design.

Evolution is the change in the characteristics of a species over several generations and relies on the process of natural selection. Charles Darwin's theory of evolution states that evolution happens by natural selection. Evolutionary theory highlights the adaptive value of within species variability. Evolutionary theory covers applications of natural selection to the ecological behavior of individuals. Evolutionary theory in biology is the change in the characteristics of a species over several generations and relies on the process of natural selection. Natural selection is a central component of modern evolutionary theory, which in turn is the unifying theme of all. Theory of evolution by natural selection was such a powerful idea in explaining the evolution of life that it became established as a scientific theory. Scientific organizations unequivocally endorse the teaching of evolution as a well- documented and scientifically important theory. Biologists continue to discuss and debate the mechanisms, patterns, and details of evolution, but the certainty of evolution is not in question. Nothing in biology makes sense except in the light of evolution. But there is a long history of creationist attack on evolution.

The general conclusion of science is that all life evolved from a last universal ancestor from bacteria-like tiny cells to animals, humans, and plants. The accepted main mechanism is that of random genetic variation, followed by natural selection. Theory of evolution is based on many observations and experiments, by which scientific confirmable evidence. Evolution is understood to the result of an unguided, unplanned process of random variation and natural selection. Mainstream scientists generally reject Creationism. But scientists are often challenged about their belief in evolution. Many Creationists try to convince people that evolution is more of a faith-based belief system than real science. One example of the scientific evidence for Creationists is the sudden appearance of complex fossilized life in the fossil record and the systematic gaps between fossilized kinds in that record. The inference from this evidence seemingly that life was created and did not evolve. The mathematical possibility that random mutation and natural selection ultimately produced complex living kinds from simple kind is infinitesimally small even after many billions of years.

The debate of the world's origins is one that has been around for many years. Two very popular views are-the views of Evolutionists and Creationists. Both are complex and have valid points, but only one can be true. The use of the natural selection was the guiding principle by which all living things came into existence by evolution theory. The Creationists view is that there is a divine Creator that had a plan in place to make the universe. Intelligent Design theory holds that certain features of the universe and of living things are best explained by an Intelligent Cause rather than undirected process such as natural selection. Scientific explanations do not go far enough in explaining the apparent design in nature. The Intelligent Cause is often assumed to be God. Intelligent does claim to explain the natural world, but is dramatically short in the explanations it offers. Creationists are often asked to propose a theory of Creation- why the particular life forms that exist were created. It states that life was gifted by a supernatural power that is invisible, but is responsible for the creation of life. God is the eternal and ultimate cause of existence, and he is the first principle.

Creationism and Evolution are different world pictures, in conceptually, socially, and pedagogically. The most common form of creationism today rejects not just evolution but much of geology, cosmology, and other sciences. Creationists are strongly opposed to a world created by evolution, particularly to a world as described by Charles Darwin. Biological evolution is genetic change in a population from one generation to another. Creationists argue that at best evolution is only a theory and even physics disproves evolution. But creation science is never truly open to check. There is no scientific to support any creationist theories, whereas nothing in biology makes sense except in the light of evolution. But there is a widespread distrust of evolutionary science. An understanding of scientific epistemology and methodology is important because it delineates useful boundaries for the creation-evolution debates. However, Naturalism is a metaphysical doctrine, which means simply that it states a particular view of what is ultimately real and unreal. According to Naturalism, what is ultimately real is nature, which consists of the fundamental particles that make up what we call matter and energy.

Creationism is an attack on the materialistic basis of science. Creationism itself is not a unified movement; its various incarnations encompass a gamut of philosophical positions; including Intelligent Design. Intelligent Design and Creationism do not just limit themselves to refuting the theory of evolution. Intelligent Design is attracting some serious attention. The argument from design for the existence of God, is based on the complex organization of living things. Humans as well as all sorts of organisms, in their relations to one another, and to their environment, could not have come about by chance, but rather manifest to have been designed for serving certain functions and for certain ways of life. Only an omnipotent and omnipresent Creator could account for the perfection and functional design of organisms. Creationism and Intelligent Design are terms used to describe Supernatural explanations for the origin of life, and the diversity of species on this planet. The teaching of supernatural accounts as alternative to evolution, and includes the concept of Intelligent Design. Creationists and Intelligent Design enthusiasts are committed to some form of non-naturalist origins.

Intelligent Design does not name a specific entity as the originator of life, but it does propose that life was initiated by a master intellect that operates outside of the known natural laws. Positioning Intelligent Design as scientific theory is in appropriate, because it lacks empirical support and portions of it are un-testable. Many scientists argue that Creationist and Intelligent Design arguments fail to meet commonly accepted criteria of science, invoking non-naturalistic explanations and failing to develop testable or falsifiable hypotheses. Science does not entertain the action of God as a viable hypothesis for any biological process. Scientists disregard non-naturalistic explanations because they cannot be defended by testable and verifiable evidence. Science tells us what the universe is like, not what meaning we may find in our existence. Science has not proven the non-existence of God; as it lies beyond the boundaries of science. Science simply looks for natural phenomena employing a methodological naturalism. Scientific data, strictly interpreted, describe the natural realms- patterns and processes.

Irreducible Complexity theory provides convincing evidence of the need for an Intelligent Design. The Irreducible Complexity means a single system composed of several well-matched, interacting parts that contribute to the basic function, wherein the removal of any one of the parts causes the system to effectively cease functioning. It argues that Darwinian evolution cannot produce such systems, by gradual modification of precursor systems, because any precursor to an Irreducibly Complex system that is missing a part is by definition nonfunctional. Intelligent Design theorists claim that certain biological features cannot be explained by evolution even in principle. Smart thinkers feel that scientifically Creationism is worthless, philosophically it is confused, and theologically it is blinkered beyond repair. A number of well-known philosophers have begun to make encouraging sounds about Intelligent Design. The theory of evolution encompasses the well established scientific view that organic life on our planet has changed over long periods of time, and continues by a process known as natural selection. Intelligent Design advocates feel the information content of DNA cannot arise by naturalistic means.ss

The Creation-Evolution controversy is obviously an interaction between science and religion. The biological community considers the evidence for evolution is overwhelming and beyond doubt. Much of the literature of the Creation-Evolution debates focuses on the details and interpretation of scientific data. There seems to be misunderstandings about the basic nature of scientific knowledge and scientific process. The central questions are how we can best come to know in a reliable and rigorous way the nature of the universe in which we exist, and whether there is room for the supernatural in scientific explanations. Intelligent Design proponents claim that methodological naturalism is philosophically inadequate because it restricts the search for truth. Some philosophers are sympathetic towards the replacement of methodological naturalism with a metaphysical neutral epistemology to make science more rational and objective. Creationists say some larger force or intelligence, or what some call agent causation seems like a viable cause for creating information systems such as the coding of DNA. Yet, it is possible to believe strongly in the theory of evolution and accept every scientific fact demonstrated.

Creationists argue that mutation and natural selection apparently could not have brought about evolution of present living kinds from a simple first organism. According to Hindu Creationism of species on earth including humans have devolved or come down from a high state of pure consciousness- Brahman. The existence of human soul, according to most religious beliefs, is a spiritual and immortal entity distinct from the human body and which is created by God for each human being individually- thus also stands outside the mechanism of evolution. The laws of nature are so finely tuned that atoms, stars, and galaxies self-assemble out of the fundamental particles produced by the Big Bang. Physicists say that laws of nature are extremely fine-tuned. In fact, chemistry and complexity would not be possible without special laws of nature. Even evolution would not be possible without them as well. Cosmologists agree that our universe appears exquisitely fine-tuned for life and extremely unlikely to have arisen by brute chance. The designer would most reasonably be God. Positioning God is not solution; because the question that immediately arises seems to be who created God?

Evolutionary theory is the unifying theme of all modern biology. Species evolve over time through the mechanism of natural selection acting on variability produced by genetic diversity and mutation. Evolutionary theory unifies the life and earth sciences with physics and chemistry. Scientists express that Intelligent Design is not a scientific concept, but it act as the wedge that splits science wide apart. What must be realized before examining these two theories is the fact that both require faith, which means neither can be proven right completely. Society has always argued about which theory is true. There are arguments on what is right and wrong, when it comes to Creationism. Creationism and Evolution is a heavily discussed topic all over the world sparking controversy on every turn. There is a lack of evidence to support the fundamental idea of

Creationism. Evolutionists strongly believe man and beast are blood related because they evolved from the same ancestor and have developed naturally. The universe is composed of energy, matter, time, and space, which is organized into durable systems that are sustaining by immutable sets of governing principles.

The theological concept allows evolution as a means of God's natural revelation. However, Intelligent Design advocates claim that God's appearance in nature can only come through intensive supernatural interventions rather than natural phenomena. They feel that there is no reason for God to act outside of the natural laws through which God creates. But Intelligent Design Advocates worry that evolution is too random and unguided to permit faith in God. Scientists point out that current scientific understanding of the universe and evolutionary processes involve more than random chance. The universe contains a chemical abundance of the elements necessary for life. Science researchers argue that the most prominent understanding of evolution today finds inherent trends toward complexity, and even intelligence and self-awareness, in the process of natural selection. The critics of Intelligent Design say that it is invalid science, narrowly conceived philosophy, and bad theology. Intelligent Design advocates challenge that there is a overwhelming evidence for designer and they point out their opponents are atheists entrenched in scientific orthodoxy.

Whether we realize it or not, we detect design constantly in our everyday lives. Everyday life as we know it would be nearly impossible without the ability to use information. Information forms the chemical blueprint for all living organisms governing the assembly, structure, and function at essential all levels of cells. Studies of the cell reveal vast quantities of information in our DNA stored biochemically. The coding regions of DNA possess high information content- where information content means precisely complexity and specificity. Many authorities have recognized the computer-like information processing of the cell and the computer-like information-rich properties of DNA"S language-based code. Life is a DNA software system containing digital code and the cell is a biological machine. Intelligent Design has its roots in information theory. Life is fundamentally based upon a vast amount of complex, and specified - information that is encoded in a biological language. The central theories of Intelligent Design show why the presence of information and meaningful complexity requires the involvement of intelligence. Intelligent Design finds in nature reliable indications of the prior action of intelligence.

Darwin accounted for design without designer or without any need to resort to external agent. Darwin argued that this appearance of design could be more simply explained as the product of a purely undirected mechanism- namely natural selection and random variation. Modern neo-Darwinists have similarly asserted that the undetected process of natural selection and random mutation produced the intricate designed-like structure in living systems. They affirm that natural selection can mimic the powers of a designing intelligence without itself being guided by an intelligent agent. Thus, living organisms may look designed and that appearance is illusory. Some scientists argue that cosmological fine-tuning does not provide evidence for of Intelligent Design, but instead the illusion of Intelligent Design. But Intelligent Design theorists argue that there are specific features of life and the universe can affirm the existence of God, they deny that God's designing activity is detectable in the natural world. The fine-tuning of laws of physics and chemistry are extreme example that allows life exist.

Intelligent Design advocates offer several arguments to cast doubt on evolutionary theory and try to promote intelligent design in its place. Their most common claim is that some biological systems, particularly in the cellular level appear to be irreducibly complex which means they must be fully formed with all their parts in place before they can serve their function. Such systems could not have evolved gradually. Because earlier nonfunctional stag been favored by would not have offered any advantages and therefore could not have been favored by natural selection. They feel the systems could have been constructed all at once by an intelligent, which can prove that only intelligence not evolutionary processes, which could have known how to arrange the process. They further say that mathematics can prove that only intelligence not evolutionary process could have produced the organized and complex phenomena, which we find in biological world today. But scientists object that the concept of irreducible complexity relies upon a mischaracterization of biological mutation as a relatively linear process involving only the addition of more and more parts, rather than a dynamic process.

Scientists object that the concept of irreducible complexity relies upon a mischaracterization of biological mutations as a relatively linear process involving only the addition of more and more parts, rather than a dynamic process that can also reshape, rearrange, or fundamentally alter existing elements and features. Systems that must be fully formed to serve their current function could have developed from earlier forms that served a different function, or could be significantly reorganized versions of an earlier form that served the same function. A loose combination of Intelligent Design arguments can stand toe-to-toe with evolution as a competing theory, even though it does not meet technical definitions and standards set by science, Further, scientists feel Intelligent Design is a less comprehensive alternative to the evolutionary theory. The scientific flimsiness of Intelligent Design along with creative science has led many critics to dismiss it. Critics of Intelligent Design have argued that although we may observe through experience that various structures are always made by intelligence, we can still argue that they were constructed by natural selection.

Creationists simply believe that God made current life forms from scratch. Evolutionists attach a great role to the evolutionary engine posited by Darwin: natural selection alone could have driven life all the way from pond scrum to us. Intelligent Design proponents believe only that the complexity of the natural world could not have occurred by chance- some intelligent entity must have created the complexity. The basic issue is that either the existence of intelligent humanity, or the mere complexity of living organisms can be accounted for by natural causes, or they point to an intelligent designer independent of nature. The irony today is that just as Intelligent Design has gained some scientific traction, discoveries in biology have weakened many of its most potent points. Irreducibly complex molecular structures point to an Intelligent Designer, but such structures have never been demonstrated to evolve by numerous, successive, slight modifications. Darwinism says that random mutation and natural selection explains almost everything we observe in living organisms. We might discover that all explanations depend on some ultimate reality that is unexplained.

CHAPTER THREE

What is the Real Nature of Time?

Time is a curious and slippery concept which continues to defy definitive explanation despite hundreds, even thousands of years of trying. Time's essence dominates our existence and yet escapes our comprehension. Time is enigmatic and ineffable. Our daily lives are completely wrapped around the onward rush of time. We work on time management- the process of organizing and planning how to divide our time between different activities. In physics time plays a major role in the measurement of motion and force. We feel the passage time in the present, our interaction with the world in the present creates our memories, which are then relegated to the past as we venture into the unknown future. But generations of physicists have claimed that time is an illusion. We can actually accept the virtual nature of time, because we have no direct sensory mechanism to sense or perceive time. Our languages are biased towards time paradigm of time, and words are inherently biased to operate with the time as a line- frame. Time perception is, therefore, a factor of awareness and attention- in one word consciousness. Theoretical physicists say that time is an illusion.

Time is not only at the heart of the way we organize life, but the way we experience it. We experience time on earth as the changes of seasons, as the process of ageing, and hours on a daily clock is surely pronounced for many of us at this time. Humans have been pondering this question of time for at least 2500 years, and so far no one has come up with an answer satisfactorily. The trouble with time started a century ago, when Einstein's special and general theories of relativity demolished the idea that time as a universal constant. Time does not need us to exist as it existed before our species was born, and it will exist after our species is dead. We create time as a way to give order to our experience. Time is the yardstick through which we measure existence: we humans are creatures of time. Time as a line we travel along is simply a mathematical abstraction that helps us understand the world. Time perception has to do with the intensity of the observer's experience. No man can evee step into the same river twice: everything keeps changing even we cannot notice the slight changes. Time exists as a word, a notion in a people's language.

Bhagavatam states that time is the impersonal feature of God. The sun, the moon, and the stars come up in the sky at same spot, same location, every single day for millions of years. When everything is the same as it was millions of years ago, then where is time, and how has it moved forward? Even the words like days and years were also invented by humans in order to successfully navigate the events back and forth, but in reality there are no days and years either. To a fruit-fly there is no such thing as time: there is only the present moment. The concept of time is simply an illusion made up of human memories, everything that has ever been and ever will be is happening right now. This is the important theory according to a group of esteemed scientists, who aim to solve one of the universe's mysteries. Time is a psychological construct because we do not see, touch, smell, or think about time. Time is not a thing and as such it does not exist. But movement does exist, and inorder to measure it, we need what we call time. We feel as if we sweep through time on the knife-edge between the fixed past and the open future, that edge- the present moment strangely appears nowhere in the existing laws of physics.

Upanishads say time is embedded in our mind and to unravel the mystery of the universe, our own mind is the key. Time is the one thing that has anchored humanity, since the dawn of civilization. Our entire lives are built around time. In theory of relativity time is woven together with the three dimensions of space, forming a bendy, fourth dimension of space. Time is merely the way we calibrate and reference change. Time is not something having its own independent existence, but is a creation of living minds. Time is only fundamental to compare changes, while engrossed in an interesting activity. Time flies, but in meditation it stands still. Time is an illusion born of our incomplete knowledge; it is not something that exists objectively. There is change in the universe, and there is our perception of it. All of physically existence .is contained within now. The future and past are not places that physically exist. The future is a prediction, and the past is memory. Unfortunately, our brains seem hard-wired to conceive of time as something that advances in a line. The empirical fact is that time advances at different rates for different observers.

Maharshi Aurobindo says time without us or an individual is absurd, because for existence per se has no time value. Time is just a measure, and just movement. If nothing were to move, there would be no time. If we were to stop thinking, we would cease to experience change as a linear progression, and the reality of timelessness would be obvious to us. Time is only the fourth space dimension in the space-time continuum. The law of quantum mechanics describing the probabilistic behavior of particles at the quantum scale shows that irreversible changes occur which differ the past from the future. Developments in physics suggest the non-existence of time is an open possibility, and one that we should take seriously. Physics equations consist of functions that tell us what happens at certain times, by the idea of viewing time as a fourth dimension. As we travel at the speed of light, and light reaches us, for us everything appears still; there does not appear to be any time. Time as we observe it is purely a way of conceptualizing the relative rates of change of different processes. Although some physicists propose tah time does not exist, time perception- our own sense of time does.

Buddhism says that we believe time's existence due to things we feel. The world without time is nothing more than a network of interconnected events. If time is an illusion, where would that leave the possibility of time travel? One of the remarkable aspects of loop quantum gravity is that it appears to eliminate time entirely. However, our own entire lives are built around time. Time may not exist as a critical component of the laws of physics, but it could still exist in our world as a social construct. Space-time does not evolve, it simply exists. There is nothing in the laws of physics to state that it should move in the forward direction as we know. Everything is real in the sense that the past, and even the future, are still there in space-time making everything equally important as the present. One consequence is that the past, present, and future are not absolutes. Time may not exist as a critical component of the laws of physics, but it could still exist in our world as a social construct. Time as we know it with the metrics we use to measure it do not exist in the world of physics, because the fundamental equations of the world do not have a variable for time.

Augustine Hippo, an ancient philosopher says that time cannot be satisfactorily described using one single definition. Time, in it fullness is unique and cannot be defined by any other concept. This view is now generally accepted among philosophers of time. Plato, Greek philosopher expresses that time is the moving image of eternity, while another Greek Philosopher Aristotle suggests that time is the number of motion with respect to earlier and later. Many philosophical and religious schools have assumed that no beginning or end can be attributed to time. Indian thought seems to tell that the universe is largely concerned as undergoing repeated creation and dissolution. Most of the medieval philosophers felt that time is considered to be relational; that is there is only be time in relation to a world of events. In Taoism time is considered to be limitless and therefore not a dominant force in our lives. Buddhism teaches about the impermanent nature of life and believes human experience of time to be an illusion. The source of time is being, and the source of being is time. But, when we see our lives from the broad view of time, we see that oue lives are not something separate from time.

The ancient Egyptians, for example, built tall obelisks that would cast shadows to help divide the day into sections: these obelisks worked in much the same way as sundials. Other devices developed over time included hourglasses and water clock, which relied upon the time it would take a particular substance such as sand or water to move from one part of container to another. Ancient cultures developed and created calendars to measure, mark, and keep track of time. Celestial bodies-the sun, moon, planets and stars have provided reference for measuring the passage of time throughout our existence. Sometime in the fifteenth century clockmakers started to use tightly coiled blades of metal-springs- to power their time pieces, instead of gravity. In the middle of the seventeenth century, pendulum clocks with an anchor escapement were made with accuracy. In the eighteenth century, the clock emerged as a scientific instrument in its own right. Measuring time also changed in the twentieth century through the development of the atomic clock and the accuracy of the atomic time continues to be refined through research. Thus time has been the product of the human imagination.

Time is an amount of successive moments. We say time waits for no one, and there is a place and time for everything. We cannot reach out and grasp it, nor can we watch it pass, yet time exists anyway. The laws of physics do not explain why time always points to the future. Time is not something that exists apart from the universe, and there is no clock ticking outside the cosmos. The problem, in brief, is that time may not exist at the most fundamental level of physical reality. If there are no changes happening around us or inside us, we would then feel we are frozen in time. Motion is of course the combination of time and distance: it is impossible to try and measure time without motion. What we call time is something that the human mind imposes on to its experience. Every watch has moving parts, as do sundials, hourglasses, and timekeeping candles. Without change, nothing would exist and all existence would completely cease to be. What science measures and calls time really exists or not, will not change as we experience it. Time is a dream we have created to help us understand and give meaning to our existence. Time in the sense of duration or the interval between two points is a derived parameter.

Time is not like space, as we are free to move around it. Time cannot be controlled and we cannot move around it. Whatever time is, for reasons over which we still puzzle, we can measure that passage of time. We measure time, keep time, meet and greet in time. Clock time is a human construct, something we came up with to make sense of what we feel deep inside about the nature of time: that time is a measure of change. Physics is not adding to the common sense understanding of time. There has been ongoing physical change, from the sub-atomic particles to that of cosmic expansion- we perceive this physical change as the passage of time. Relativity theory says that what we perceive as time is just a fourth dimension of space, and both quantities are measured. Space-time is a different concept, where space means all that is and time means the perception of motion created to give context to the present. Everything in the universe has an underlying rate of change. All of the confusing things about time disappear, when we conceptualize reality as an ever changing now: the universe will make a lot more sense. Our naïve conception of time's flow does not correspond to physical reality.

There is no such thing as time- there is only the present moment. Without consciousness, time perception is impossible. We think of time having three parts: past, present, and future, which are only a collection of thoughts about it-memory. The more conscious we are of our surroundings and our minds, the more changes we notice. In consciousness time is always now. The universe exists only in the now. Physicists affirm that reality is just a complex network of events onto which we project sequences of past, present, and future. The whole universe obeys the laws of quantum mechanics and thermodynamics, out of which time emerges. The discovery that time does not exist may have no direct impact on our lives, even while it propels physics into a new era. Perhaps what physics is telling us is that causation and not time is the basic feature of the universe. The apparent existence of time in our perceptions and in physical descriptions comes not from knowledge, but from ignorance. It looks that we need a new physical theory to explain the universe, and this theory might not feature time. Time has no meaning in itself, unless we choose to give it significance.

An understanding of time is one of the most deeply held assumptions of human culture. We do have a strong sense of time that plays a crucial role in every conscious decision we make in our lives. We can argue that the reason for the existence of time is our knowledge of our finite life-span. In physics we find Newton's constant time and Einstein's malleable time. The difference between these notions of time is indicative of its unreal nature. Time is unreal in the same way as mathematics is unreal: they are both products of our intellect. Philosophically they can be thought as formal language. Hardly anything else determines our lives and also our attitudes towards life as much as thing we call time. One of the beliefs shared by global religious traditions is the importance of sacred time and sacred place. Both sacred place and time are vital to authentic worship and can function to draw us into the experience of God. Time and consciousness have long shared objective philosophy and social thought. Timelessness is a state of eternal existence believed in some religions to characterize the after- life. But time remains basically as an illusion created by the mind to aid in our sense of temporal presence in the vast space.

The concept of time as we know it, is not a fundamental reality. Time is a story we are always telling ourselves in the present tense, individually and together. It makes sense that time is an arbitrary construct- it is our way of making sense of growing up and growing old, while the world changes around us. The eternal time is reality- it has beginning and an end is illusion. In the eternal time there is only one tense- the present. Time is basically a concept of consciousness. We can only experience the present moment. It is the invisible presence of time that governs our life. Time is actually discrete, and yet, it appears perfectly continuous. When I say that I was born on 6 April 1943, in reality it was a unique place in spae that will never be touched again. Time moves differently for someone below sea level than for someone situated on the highest peaks on the planet. Time can be considered as the fourth dimension of reality, used to describe events in three dimension space. Physicists define time as the progression of events from the past to the present into the future. The dimensions of is of clear importance for adaptation and orientation in the physical and social environment.

Time is familiar to everyone, yet it is hard to define and understand. Science, philosophy, and religion have different definitions of time, but the system of measuring it is relatively consistent. We can never experience time directly, and time as a discrete force of nature is an illusion. It is not something we can see, touch, smell, or taste, but we can only measure its passage. The question of why time is irreversible is one of the biggest questions unsolved in science. Humans are aware of the passage of time and are highly influenced by attention demands during a target interval. Psychological time is a complex notion reflected in many types of experiences. Time seems to slow down during emergencies or danger. Older people seem to perceive time as moving faster than when they were younger. An atomic clock aboard of a jet airplane will tick at a slower rate than one on the ground. In mathematics time can be defined as an ongoing and continuous sequence of events that occur in succession, from past through the present, and into the future. Some philosophers and scientists have argued that we experience as time is just n illusion, an artifact of our consciousness.

Brahmanical tradition has maintained that the ultimate being is timeless. If we see the intersection of time and space, we experience complete freedom of being. This state of existence is completely beyond any idea of time, space or being. In that liberated state, we can see fundamental truth and the phenomenal world simultaneously. In the common sense, time seems to be something separate from beings. The essence of relativity is that there is no absolute time; no absolute space. The time we know since we learned to tell time on a clock, which seems to disappear when we learn physics. In the block universe, as supported by Einstein's theory of relativity, as a four dimensional space-time structure where time is like space, in that every event has its own coordinate or address in space-time. No observer has knowledge of distant event or the simultaneous different events, until they are unambiguously in that observer's past. The block universe theory posits that our universe might be a giant four-dimensional block of space-time, containing all the things that ever happened and will happen in our traditional perception of time.

Time is nature's way of keeping everything from happening at once. Time has been characterized as a structural coordinator of reality. Time is also labeled as the currency of life. Time heals all wounds. Time is free, but it is priceless. Time is actually the most precious and valuable resource in our lives. Every cell in our body has its very own clock. Plant clocks prompt their leaves to open during the day and close during the night. But science tells us that all moments that exist are relative to each other within a three spatial dimensions and a single time dimension. The block universe theory is also termed by some scientists as Eternalism, because it describes how the past, present, and future all co-exist. From the block universe point of view, time or temporal relations of earlier than and later than exist. These relations hold regardless of where anyone is located. Just on this model now picks out whatever time we happen to be located at. Past picks out any time or events at those times that are earlier than our location, while future picks out any times or events that are later than our location. Time is the continuous sequence of events that take place in irreversible succession.

A person lives in the world within a time and everything around him changes with the passage of time. Time is related to all that is going on around us, so we often hear about the time dimension, because time is essential for life. In time we know the history of of the events that are taking place around us, and it helps us in documenting these events. Time also matters in determining the ages of antiquities and artifacts. We find that all worship has a close relationship with time. Time is one of the most mysterious forces in the universe. Time acts as both a teacher and a healer. Time teaches us the value of life and makes us feel happy to be alive. Nobody knows how much time they have. We can die at any age and for any reason. We schedule the day according to the concerns that we may be experiencing. Time is the one thing that we can never get back once it has gone, and will never return. Human life is measured by the time, and it plays such a significant role in human life. However, time is the construct of human mind. Time seems to follow a universal rhythm, but it does not. Time cannot be treated in isolation from space.

Time is something we deal with everyday, and something everyone thinks they understand. Time is only a reflection of change. Time is used to quantify or measure or compare the duration of events or intervals between them, and even sequence of events. Our brain constructs a sense of time as if it is flowing. WE give false import to this present moment, by the human perspective of time flows with a direction. The three basic properties of time come not from the physical world but from our mental states. For many physicists while we experience time as psychologically real, time is not fundamentally real. Time is a prime conflict between relativity and quantum mechanics. From moment to moment, all sentient beings exist together as a completely independent moment of time. But the total picture of time cannot be understood only in terms of being and time; it must also be understood in terms of space. Space is the vast expanse of the universe where everything exists. All sentient beings are dynamically interconnected and interpenetrated by the energy of timelessness. Whether time turns out to be a mere illusion or an undeniable physical fact of the universe only time will tell.

CHAPTER FOUR

Where did Mathematics Come From?

Mathematics is one of the most important fields of science. Mathematics is a philosophical question as old as philosophy itself: was the system of mathematics discovered or invented? Mathematics is involved in almost everything we do today whether we know about it or not. One cannot deny the tremendous influence that the study of mathematics has had on modern civilization. It is largely to thank for our mobile phones, computers, most of our methods of transport, and even our buildings. Mathematics describes the real world with remarkable precision. With its abstract language, mathematics is suitable to describe everything using universal concepts that can be easily understood by everyone. Mathematics is a field of science that deals with the logic of reason, quantity, arrangement, sequence, and shape. The basic philosophical question in mathematics is ontological, and is concerned with existence. Mathematics and language, both are ways of describing the current condition of nature. Both have hardwired sections in our brain, and both are found at every level of every nervous system. Mathematics as a pure subject is more aligned with philosophy rather than science.

Mathematics is so vast and is involved in so many types of studies, we cannot really say that who has invented I, as it is developed over thousands of year with many savvy individual contribution. There are traces of mathematics in the history of almost every civilization. Many people argue that mathematics existed way before we found it and had a use for it. Many historians suggest that the early adopters of mathematics were the early civilizations of Greece, India, China, Egypt, and Mesopotamia. Some of the basic functions such as addition, subtraction, and division have been used for over 4000 years. With wide understanding and acceptance of the basic mathematics, the science evolved and grew over time, with the development of geometry, algebra, calculus, trigonometry, and even probability. The miracle of the language of mathematics to formulation of the laws of physics is a wonderful gift. The uncanny ability of mathematics is not only to describe and explain, but to predict phenomena in the physical world is more astonishing. Now the real question is, did we arbitrarily create fundamental mathematics and logical truths, or did we find them in embedded in nature?

Mathematics reveals the secret of nature. Platonic school holds that mathematical objects are said to exist. Logic school seeks to reduce all of mathematics to logical thought, which means mathematics is purely an intellectual exercise. Intuition school seeks to explain that mathematics is activity of mental construction-natural numbers, real numbers, proofs, and theorems. Fiction school tells that the reality that we think is being described by mathematics in nothing more than fiction-there is an underlying reality which we know nothing about. Using mathematical things about the world and what not were able to be explained. According to realism, mathematics exists objectively and independent of human thought. Mathematical concepts are disembodied in the universe and available for us to uncover and bring into practical use. Though mathematical objects obey sets of rules, anti-realism argues that they are rules ultimately created by us, maintained by us and often revalued by us. Zero is the first whole number in our number system, which has changed the way we perceive Mathematics and science. The idea of zero was developed into a numeral in ancient India, arisen from the philosophy of emptiness.

Mathematics is the queen of all sciences. There are an infinite number of primes or truths about reality that were held even before mathematicians knew about them. Humans invent mathematical concepts by way of abstracting elements from the world around them. Some believe that mathematics exists within us, and that the objects of mathematics were therefore our creation. In 500 BC, we saw the development of Roman numerals, which are still used to represent numbers. Scientists believe that thousands of years ago, basic mathematical functions like addition and subtraction might have appeared at the same time, but in different places like India, Egypt, and Mesopotamia. Ancient Greek thinker and mathematician Plato believed that mathematical entities are abstract and exist independently in their world, outside of space and time. Many mathematicians have discovered many eternal truths, independent of the mind that found them. For example, there is no highest prime number, and the number pi when expressed in decimals can go forever. We can find mathematics manifesting itself in nature in the Golden Ratio. The golden ration describes the most predictable patterns in the universe.

Mathematics is the hidden secret to our understanding of the world. Mathematics is a discovery that mathematicians made using techniques invented by them. Mathematicians go on to discover the connections among the concepts based on their perceptions and mental pictures. Some philosophers think that mathematics existed independent of our thoughts. Advanced math dates back to Greece over 2500 years ago, when mathematician Pythagoras proffered his famous equation. The golden ratio describes everything from atoms, the shapes of a hurricane, the face, and the human body. Humans have come to understand the working of the universe simply by observing the patterns that appear in nature. Geometry and arithmetic were developed due to our ability to observe and distinguish between shapes like circles and triangles, as well as to differentiate between straight and curved lines. Mathematicians invent mathematics all the time, formalizing observations they make about the mathematical world into rules. Mathematics developed slowly over thousands of years with the help of thousands of people. Pi is a mathematical mystery that has captivated people for thousands of years.

Mathematics is the language of the universe. Pythagoras theorem- the square of t he hypotheses of a right angle triangle is equal to the sum of the squares of the other two sides is true for all right angled triangles is a discovery. Luckily for mathematicians and physicists alike, universal laws appear to govern our cosmos. The tale of mathematics is as old as the humanity. It has evolved from simple math, like counting to an intricate study of abstract concepts that we know today. Later, more mathematicians started working on expanding their understanding of mathematics. The golden ration is also known as divine proportion, and has a value of about 1.618. The absolutist view of mathematics sees it as universal, objective, and certain, with mathematical truths being discovered through the intuition of the mathematician and then being established by proof. According to social constructivism, mathematics is more than a collection of subjective beliefs, but less than a body of absolute objective knowledge, floating above all human activity. Mathematics is cultural knowledge, like the rest of human knowledge. Mathematics is surely the greatest of many inventions.

Mathematics is the historical artifact. Showing it is true, however, requires the invention of a proof, and over the centuries mathematicians have devised different techniques capable of proving the theorem. But, mathematics would not work at all were there no universal features to be discovered. There were moments in the past when people discovered something that already existed, and other times when people thought they invented equations and methods to write something that was simply going in their minds. The golden ration was derived from Fibonacci sequence, named after the Italian mathematician Leonardo Fibonacci. Flower petals also follow the Fibonacci sequence. If we observe, the number of petals in a flower will be either one of the following 3, 5, 8, 13, 21, 34, or 55. For example a lily has 3 petals, cosmos has 8 petals, corn marigold has 13 petals, and daisy has 21 petals. This supports the argument that mathematical functions existed in nature, and we did was to discover them. The absolute nature of mathematics is universal, objective, and certain, with mathematical truths being discovered through the intuition of the mathematician and then being established by proof.

Mathematics deals with secret universe of patterns, beauty, and interconnectedness. Mathematics is unreasonably effective in two distinct ways of active and passive. Some believe that mathematics is out there, simply waiting to be discovered. For hundreds of years, the Fibonacci sequence has fascinated many mathematicians, scientists, and artists. Some people oppose the idea that mathematics was discovered. They consider mathematics to be a human invention designed in a way that suitably describes the physical world. As a field of study, mathematics often enjoys a distinctive status among the sciences. Mathematics comprises a priori knowledge that deals with truths that are true by virtue of necessity. The uncanny ability of mathematics is not only to describe and explain, but to predict phenomena in the physical world. How is it possible that all the phenomena observed in the classical electricity and magnetism can be explained by means of just four mathematical equations. The mathematical entities, relations, and equations used in these laws were developed for a specific application. Where do numbers come from is a speculation- no one knows.

Mathematics and its structures are innate to nature. Even if the universe were to disappear tomorrow, the eternal mathematical truths would still remain. And over time humans in various cultures have noticed patterns that emerge in numbers, and developed mathematical systems around them. The number system used in ancient India had zero. Much of our mathematics is based on one system called base 10, which works on patterns of one to ten, and called the decimal system. The Fibonacci sequence can be seen in various items around us, including seashells, animals, pyramids, and other unexpected places. The successes that have been achieved from the mathematics of the cosmos down to electronics devices at the micro-scale, are significant. Einstein remarked- how can it be that mathematics, being after all a product of human thought which is independent of experience, is so admirably appropriate to the objects of reality. The reason mathematics is the natural language of science, is that the universe is underpinned by the same order. The structures of Mathematics are intrinsic to nature. It is up to us to discover mathematics and its workings. Mathematics at times seems to be a human construct.

Mathematics is the centre of our modern world. Some people argue that mathematics was not an invention, but discovery. The idea behind it is that mathematics exists in the mind of God or the Platonic world of ideas, and all we do is discover it. Mathematics is admirably suited describing the physical world, and that is why we invented it to do just that. Mathematics is not discovered, but it is invented-this is the non- Platonist position. Mathematical - Platonists eagerly point out that elegant patterns are common in nature, and that mathematicians clearly discover rather than invent. Philosophers of mathematics have tended to think mathematics as being with us, while other philosophers have thought of mathematics as being outside us, as existing independently of our own thoughts and feelings. Now, one might be tempted to adopt an intermediate position. Could the objects of mathematics be somehow both partly invention and partly discovered. Mathematics is a language, and we can describe things through it. Some view that mathematicians invented the axioms and the rules of operation, the rest are discovered. Mathematics is embedded in the world, and figures in paintings and perspectives.

Mathematics is the language of science and has enabled mankind to make extraordinary technological advances. Have created mathematics to understand the universe, or it is the native language of the universe itself- something made by man and something that already existed before man. Pythagorean of the fifth century believed that numbers were both universal principles and living entities, with each number its eccentric nature. Plato asserted that independent of human existence, numbers were as real as the universe itself. Philosophers of our time have thought about mathematics as being in the existence of our thinking already existing before mankind. The foundation of mathematics is the number system, which could be described as an extension of the natural numbers. In contrast, other philosophers may argue that mathematics is invented and thus a creation of our own thinking. From that follows we are indirectly thinking about ourselves; we use our thinking ability to set a framework of the universe. Yet other philosophers view that mathematics can be seen as an artifact, an axiom system peculiar to this humanity. Mathematics has become the vital backbone in understanding everything.

Mathematics manifests itself in nature and holds answers to many universal questions. Some philosophers feel that some mathematics is definitely invented, not discovered. We find some really deep, fundamental mathematical idea, and there it really looks inevitable. One idea is that mathematics is in the mind of God- it is all there, and all we do is to discover. Few mathematicians have proposed the theory that mathematical statements do not exist, no matter whether the existence of numbers is physically true or not. The reason behind this is the true values of them are just human creation. It is also proposed that mathematics is nothing but an abstract relationship implying on patterns that are discerned by brains and are built to figure out artificial order from any branch of chaos. Greek philosopher Plato argued that mathematics is a discoverable system that underlies the structure of the universe. In other words, the universe is made of mathematics, and the more we can understand this vast interplay of numbers, the more we can understand nature itself. To put it simply, mathematics exists independent of humans. There is its ever-growing application from medicine to economics.

Mathematics is the centre of our modern world. Some people argue that mathematics is not an invention, but discovery. The idea behind it is the mind of God or the Platonic world of ideas, all we do is to discover it. Mathematics is admirably suited for describing the physical world, and that is why we invented it to do just that. Mathematics is not discovered, but invented- this is the non-Platonic position. Platonists eagerly point out that elegant patterns are common in nature, and that mathematicians clearly discover rather than invent. Philosophers of mathematics have tended to think that mathematics as being with us, while other philosophers have thought of mathematics as being outside us, as existing independently of our own thoughts and feelings. Now one might be tempted to adopt an intermediate position. Could the objects of mathematics be somehow both partly invention, and partly discovered. Mathematics is a language and we can describe things with it. Some view that mathematicians invented the axioms, and the rules of operation, the rest are discoverable. Physicists say that our external physical reality is a mathematical structure.

Mathematics manifests itself in nature and holds answers to many universal questions. Many people think mathematics is a human invention. To this way of thinking, mathematics is like a language: it may describe things in the world, but it does not exist outside the minds of the people who use it. Its proponents believed reality is fundamentally mathematical. If mathematics explains so many things we see around us, then it is unlikely that mathematics is something we have created. The alternative is that mathematical facts are discovered, not just by humans, but by insects. Some believe physical reality is made of mathematical objects in the same way matter is made of atoms. In the past century physics has become more and more mathematical, turning to seemingly abstract fields of inquiry such as group theory and differential geometry in an effort to explain the physical world. As the boundary between physics and mathematics blurs, it becomes harder to say which parts of the world are physical and which parts of the world are mathematical. Few mathematicians have expressed that mathematics is just invented logic activity hardly proving any existence outside the consciousness of mankind.

Mathematics is how the universe, which is conscious, comes to know itself. Galileo said that mathematics is the language of God. In this sense we discover the hidden coding o f the universe. The opposing argument, therefore, is that mathematics is a man-made tool- an abstraction free of time and space that merely corresponds with the universe. The most common pattern in the universe is Golden Spiral. Scientists have long used mathematics to describe the physical properties of the universe. All matter is made up of particles, which have properties such as charge and spin, but these properties are purely mathematical. There is an elegant simplicity and beauty in nature revealed by mathematical patterns and shapes, which our minds have been able to figure out. The mathematical nature of the universe helps scientists in developing theory that predicts every observation or measurement in physics. The human mind is probably the most complex structure in the universe, and consciousness is probably the way information feels when it is being processed. Many great breakthroughs in physics have involved in unifying energy and matter; space and time; electricity and magnetism.

Mathematics is a useful way to think about nature and our world. Mathematical knowledge is always clear, logical, and systematic. When we use search engines like Google or buying a product online. Or watch a movie on an OTT platform, it is all mathematics. When we use a laptop to send emails there is mathematics going on inside the machine. The binary number system uses only two digits of o and 1 to express all mathematical concepts. Mathematics is central to the Internet messages and World Wide Web. Mathematics is the science of numbers and their operations, interrelations, and combinations. Science provides mathematics with interesting problem to investigate, and mathematics provides science with powerful tools to use in analyzing data. The symbolic language of mathematics has turned out to be extremely valuable for expressing scientific ideas unambiguously. Using mathematics to express ideas or to solve problems involves at least three phases: 1.it represents some aspects of things abstractly: 2. It manipulates the abstractions by rules of logic to find new relationships between them, and 3. It sees whether the new relationships say something useful about the original things.

Mathematics today is a diverse discipline that deals with data, measurement, and observation from science: with inference, deduction, and proof: and mathematical models of natural phenomena, of human behavior, and of social systems. Geometry, algebra, trigonometry, and calculus all play a crucial role in science and engineering. It employs observation, simulation, and even experimentation as means of discovering truth. In addition to theorems and theories, mathematics offers distinctive modes of thought which are both versatile and powerful. Mathematics today also involves modeling, abstraction, optimization, logical analysis, inference from data, and use of symbols. Mathematicians feel that there is something mysterious about this queen of science. There are certainly mysteries that exist in mathematics. Mathematical theorems like that of Pythagorean are seemingly discovered. We discover mathematics signature in the swirl of a nautilus shell, the whirlpool of a galaxy, and the spiral in the centre of a sunflower. Mathematics from Pythagoras to Einstein and beyond, all leading to the ultimate riddle: Is mathematics an invention or discovery?

Mathematics plays such a central role in modern culture, some basic understanding of the nature of mathematics is requisite for scientific literacy. Mathematics reveals the hidden patterns that help us understand the world around us. The essence of mathematics lies in its beauty and its intellectual challenge. Mathematics relies on both logic and creativity. As science of abstract objects mathematics relies on logic rather than on observation as its standard of truth. The alliance between science and mathematics has a long history, dating back many centuries. Mathematics provides the grammar of science-the rules for analyzing scientific ideas and data. Religious thinkers say man did not invent or create mathematics- it is God's creation. Mathematics requires order and God created that order. The absolute consistency of mathematical principles is possible because God was the inventor of that consistency. Throughout history, mathematics has offered glimpses of infinity that point man's attention to God. Some believe that mathematics was out there simply waiting to be discovered, while others believe it is instead a creation of our mind. Even today the question does not have an exact answer.

CHAPTER FIVE

Are We Alone in the Universe?

Are we alone in the universe? What at first appears to be a very modern query it is actually ancient in its origins. Is anybody out there? This question has three answers: yes, no, and don not know. Are we alone? Is a question that is both profound and eternal. The only correct answer to this question for now is that we do not know. This question is worthy of serious scientific study, and unfortunately neither of the answers feel satisfactory. This question is as old as humankind itself. Scientists are actively searching for signs of life in the universe with a belief that it is teeming with life. Ancient mythologies and contemporary science fiction have presented imaginative answers of potential for life on other planets. The answer to this scientific question affects our claim for being special in the cosmos, and we shall never know unless we search. The real tension, moving forward, drives from our own uncertainty about who we are now, what we wish to become in the future, and what we might unknowingly risk becoming in the process. Two possibilities exist: either we are alone in the universe, or we are not. Both are equally terrifying.

Today those who reject the notion of alien life are considered to be the outliners. This change in attitude occurred, perhaps none as important as the discovery that exoplanets are everywhere. Surely, there are so many opportunities for life to flourish, and therefore, the universe must be definitely teeming with extraterrestrials. Each point of light in the night sky has its own solar system of orbiting planets, and each exoplanet out there ha s a potential cradle for life. With roughly 100 billion galaxies hosting countless stars each, the universe seems like it should be teeming with life. And yet, we do not have a conclusive answer, as we have never received a credible radio signal from extraterrestrial life beyond earth. Astrobiologists say that the only correct answer to this question for now is that we do not know. But they are hopeful of the existence of extraterrestrial life with the ongoing exciting research into the origins of life. We have been fascinated by the idea life may exist elsewhere, and for over 60 years the Search for Extraterrestrial Intelligence SETI has been scanning the galaxy for messages from an alien civilization to no avail.

The possibilities of technological civilizations existing elsewhere in the universe are derived from the Fermi Paradox, and the Drake Equation. But, we have not heard any technologically advanced alien civilization. The Fermi Paradox seeks why we have not heard from anyone, while the Drake Equation attempts to spin a more optimistic approach to the conundrum. We have yet to confirm a definitive radio signal from Extraterrestrials, or develop the technology needed to explore the universe beyond the solar system with a spacecraft. However, the discovery of extraterrestrial life would be hugely exciting, but it does not change our place in the universe. Many astrophysicists say that intelligent life probably exists on distant planets, even if we cannot make contact. Recent unidentified aerial phenomena have set off another round of speculative musings on the possibility of aliens visiting our planet. We are unlikely to be able to communicate or interact with them. To understand what is out there is a timeless human drive. According to our best estimate, the universe is home to a hundred billion trillion stars- most of which have planets revolving around them.

The newly revealed trove of orbiting exoplanets greatly improves the odds of our discovering advanced extraterrestrial life. Scientific evidence from astrobiology suggests that simple life-composed of individual cells or small multi-cellular organisms are ubiquitous in the universe. Given all that real estate, it seems unlikely we are alone. The same physical rules that apply here are no different elsewhere in the universe. Despite of decades of searching, we have not detected even a single robust signal that indicates the presence of intelligent aliens. If the ingredients of life are everywhere, and there are astronomically large numbers of stars and planets, where it is possible for life to have arisen, then we would expect many instances in which intelligent aliens rose to prominence well before the advent of human life on earth. If the universe is teeming with life, then where is everybody? Sometimes we think the surest sign that intelligent life exists elsewhere in the universe is that none of it has tried to contact us. There are many theories around that attempt to offer a comforting thought about life in the universe. Of course, in the end, there is no way to really tell at this point whether or not we are alone.

Perhaps when we look at the night sky with enormous stars, we get an innate feeling that somewhere out there, the mysterious spark we call life has flickered into existence. Intuitively, we feel as if we cannot be alone. The Hubble Telescope has given us spectacular pictures from outer space, but they are not detailed enough to help us in our search for life in the trillions of galaxies across the universe. Our serious searches for any signatures of life stretch around a few thousand nearest stars, which is an extremely small search space. The probability of such a large universe featuring alien life so likely, then it stands to reason that an advanced alien civilization should have reached out by now. Assuming aliens do exist, questions such as where we should be able to find them, how we could go about looking for them, and what signs we should try looking for. It would be reasonable to infer that whatever is common here on earth is plentiful throughout the universe. And even if the life does arise elsewhere, it may be the case that it frequently fails to thrive. In all of space, as far as intelligent goes, humanity is truly alone. Astrophysicists studying the universe have revealed, our own planet as something less than special.

Some eminent astrophysicists argue that signs of intelligent extraterrestrial life have appeared in our skies. Our skies are not alive with the cross-talk of millions of interstellar communications. It seems that whenever astronomers discover anything unusual in the sky, somebody claim it must be alien. There are undeniably some UFO sightings that cannot be dismissed so easily, because alien visitation is possible. The observation of an object with a potentially strange trajectory around the sun, has reignited in the question of UFOs and the existence of intelligent alien life. UFOs are coming out of the shadows. The UFO phenomenon as we know it today is much more recent, dating to the era of powered flight. UFOs really took off during World War-2, when allied pilots from rival group of war reported seeing puzzling lights or objects in the sky. The first report of flying saucers began to surface during the summer of 1947. UFO sightings or encounters have been documented ever since the first report in 1947. For nearly 75 years people have been observing UFOs all around the world. The general public also seemed dubious that flying saucers might be the work of extraterrestrials.

For better or worse, sightings of unidentifiable flying objects have become inextricably linked to visitors from outer space. Aliens are now our default explanation for such events. UFO continues to carry with it the same association with aliens up to now. And in November 2004, several Us Navy pilots flying off the coast of San Diego reported seeing bizarre craft zooming through the sky, seemingly maneuvering in ways that exceed the limits of known technology. The UFO phenomenon has become enmeshed in a web of conspiracy theories that include accounts of abduction by aliens. These conspiracy theories have undoubtedly been fueled in part by the fact that the military has been secretly investigating UFOs for decades. UFO phenomenon warrants investigations after decades of studies and hundreds of thousands of sightings. Scientists will have to get over their reticence to engage with the UFO sightings. Scientists need to shine light on this decades-long mystery. Millions of people elsewhere in the world believe that UFOs are real, as there has been sightings of them by millions people for decades. UFOs are undeniably real; people often see things in the sky that cannot identify.

Some UFOs appear to remain stationary in winds aloft, move against the wind, maneuver abruptly, or move at considerable speed, without discernible means of propulsion. Alien abduction stories are more complicated, as they tend to involve more psychological components. The aliens are kind of keeping pace with technology. However, the extraterrestrial idea should not be dismissed or ridiculed. Science should reject potential extraterrestrial explanations because of social stigma or cultural preferences that are not conducive to the scientific method. With countless books, movies, and TV shows about UFOS, the possibility of extraterrestrial life has captured the imagination of global population for decades, and it is something that fascinates billions around the world. More than 50 percent of people around the world believe in the existence of intelligent alien civilizations in the universe, while greater than 60 percent of people believe there is some form of life on other planets. There is certainly global interest in aliens and UFOs as evidenced by surges in Google searches following major news stories. Some special investigations asks us to forget everything we have ever been told about UFOs- they are real.

Psychological research on the topic of UFOs has been surprisingly scarce. The few studies have been done in isolation from one another. The people reporting contact with aliens, known as experiencers appear to have a different psychological profile. A 2021 intelligent report describes advanced technology of UFO. Some unexplainable sightings of aircraft accelerate astonishingly faster. We need new scientific to explain the behavior of these aircraft. Science might explain the strange aerial objects, whether that science is exploited by humans or extraterrestrials. UFOs are weird-shaped- flying objects that move in shockingly abnormal ways. The majority of these sightings showed up on multiple sensors at the same time, such as special cameras and radars, and so could not be easily dismissed as purely visual anomalies. The UFOs either represent exciting extensions of known science or exploit features of exotic unknown science, such as faster-than-light travel through gravitational worm holes. We do not know that these UFOs are not manufactured objects at all, but a sort of mid-air projections that appear to be solid. The psychology of alien contact largely revolves around the concept of otherness.

Even some of the earliest recorded accounts of human existence contain references to UFOs flying across the sky or visiting earth. Humans have been reporting UFO sighting for centuries. So far, by one count more than one lakh sightings have been reported. The alien experts say there have been more than 100K recorded sightings in the past 100-plus years. Those who believe in the existence of aliens often point out to happenings and unusual locations on earth as evidence of extraterrestrial life. Most people believe that outer space aliens exist and they have visited our planet, and humans are not only civilization in the universe. Many express their curiosity into multiple UFO reports coming from the armed forces. Several prominent international instances of UFOs are available publicly. Many of these accounts are substantiated with data including detection by radar, audio recording of conversations, satellite imagery, and photos. Many experts say an inexplicable or unidentifiable phenomenon does not necessarily mean an alien or extraterrestrial phenomenon. Space experts offer an explanation for several UFOs as completely unknown flying machines with exceptional performance attributed to aliens.

Space experts further argue that there must be alternate, physics-compliant explanation for such UFOs. The governments continue to study UFOs, and attempts to capture more data about them using advanced military sensors and modern equipment. The Pentagon has been quietly investigating UFOs since 2007 after conceding there is more to them. Aliens are an escapist fantasy, but also more credible than mere fantasy. Alien dreams have always been powered by the desire foe human importance in a vast cosmos. What is unusual about the alien fantasy is that it does not place humans at the centre of grand story- in fact it displaces them. Many theories have been sought to explain these apparitions of UFOs that defy science. UFO symbolizes an uncertainty about the boundaries between scientific knowledge. The number of UFO sightings is currently flying at an-all-time high according to data crunchers. UFOs are so prevalent in both conspiracy theory, and the new age milieu in the post cold war period. There is an interest in explaining the creation of ancient sites as the work of advanced alien civilizations. The reality is that these structures were likely built by extremely advanced civilizations.

Alien life is hypothetical life that could possibly exist outside of earth, and which did not originate on Earth. Alien life could range from simple forms of life, such as bacteria and archaea, to intelligent and even sapiens beings. In theory, alien life could be much advanced than life on earth. The science of extraterrestrial life is called astrobiology. To this day, no life outside of earth has ever been found, and science does still support the idea that alien life is possible. Astronomers have been spending lots of time trying to discern traces of simple forms of life, known as bio-signatures elsewhere in the universe. If intelligent life on alien planet built a technological civilization, their techno-signatures can possibly be detected. Despite the multitude of forms of life that exist, life must be rooted in the same physical and chemical laws that underpin the universe. Even if intelligent life forms exist, their conditions may or may not make it difficult to contact us. AS stars are far away, it could take years for a message from extraterrestrial life to reach us. Finding alien life would be one of the greatest discoveries in human history.

NASA is set to venture into one of the most controversial subjects-UFOs. The space agency is forming a science team to study the mysterious sightings. One of the most interesting questions facing scientists is what would aliens look like? They imagine extraterrestrial life being based on molecules, which is not DNA, making alien life; very different from what we see on earth. But alien life may be so different from life on earth, that it could be impossible to recognize. There is no set definition of life. While most scientists focus on alien life that would be similar to earth life, there is always the chance they will be unexpectedly different. SETI- Search for Extraterrestrial Intelligence Institute aims to explore, understand, and explain the origin, and nature of life in the universe and the evolution of intelligence. But if there is life outside there, why we have not heard from them yet? Alien spaceships might be easier to spot than their home planets. Even if aliens try to speak to us through signals, we can barely comprehend. We humans have always been curious about the unknown, and finding alien life would be one of the greatest discoveries in human history.

We on the earth now possess all the technology necessary for communicating with other civilizations in the depths of space. In our present ignorance of common extraterrestrial life may actually be, any attempt to estimate the number of technical civilizations in our galaxy is necessarily unreliable. We do, however, have some relevant facts. Since intelligence and technology have a high survival value, it seems likely that primitive life forms on the planets of other stars, evolving over many billions of years, would occasionally develop intelligence, civilization, and a high technology. Scientists are still searching for signs of intelligent alien life on other planets- but how would we react towards it if we ever did make contact? In popular culture, extraterrestrials are often cast as second-class citizens or less than human. Some argue that finding life or making contact Is always going to be highly unlikely until the day we do. A clue for how we might treat aliens we ever do have contact with may lie in the rights we have afforded non-human species on our planet. The future-focused scenarios may never happen, but the whole process has value in and of itself.

Evidence for alien technology might involve subtle and very ancient physical traces. Since most stars were formed billions of years before the sun, it is conceivable that technological civilizations that emerged around them had more time to develop their science and technology. We know that any technology that goes beyond ours would help us introduce major commercial or military benefits as well as represent consumer market or on the battle field. A decisive scientific experiment of UFOs would demonstrate the power of science in answering a question that is clearly of great interest to the public. We are misled by cognitive biases, optical illusion, or in blindness to things we do not expect to see in the sky-especially UFOS. We should be open-minded to search for alien life elsewhere, and understand their technology. Apart from identifying alien spacecraft, some feel such as UFOs has national security and air safety implications. It is possible that a single UFO of extraterrestrial origin will inspire fear among humans that aliens could be malevolent. But they could be benevolent as well, when the time comes, and learn that their intent is wise and well-meaning.

There are incredible findings about aliens from 2020. Researchers have announced that they had captured a very mysterious beam of energy in the radio part of the electromagnetic spectrum. Astrobiologists anticipate alien bacteria might live in the clouds of Venus planet. Space experts consider a cigar-shaped object hurtling around through the solar system to be something propelling it. US Navy releases footage captured by pilots that showed odd wingless aircraft travelling at hypersonic speed, which looks like a bizarre alien machinery. It has been conjectured that some fraction of alien civilizations might expand beyond their home planets, and spread across the galaxy, for reasons of exploration, colonization, or something else. This really raises the possibility that our Solar System, and even earth itself may have been visited at some stage in the history. We could look for traces of such visit or sojourn. Astronomers are to detect such unmistakable signs of alien technology. Now, the time has come to search intelligent life with all the resources and technology available, and it is about to get a lot more interesting.

Any detectable techno-signature must be within the bounds of our past light cone, which is the path that light from any single event would take through space-time. We must train ourselves to look at ourselves through alien lens. Also we might be able to better understand ourselves through the eyes of perhaps creatures on other worlds. The technologies employed in the UFOs cause rapid movements, which cannot be explained by using our physics or engineering. An advanced alien civilization might be able to manipulate space-time using anti-gravity technology. Most of what Silicon Valley has created and innovated seems to be originally derived from an alien spaceship crash in Rosewell, in New Mexico. The wreckage was reverse engineered and released to select companies for further development. There are some credible, relied upon people, all saying yes- there is UFO technology, there is antigravity, there is free energy, and it is extraterrestrial in origin, and they have captured spacecraft and reverse-engineered it. If humankind would go about it peacefully, aliens would share the splendor of their technology.

It is possible that we will see more evidence as UFO sightings happen. The reality of some UFOs is beyond doubt, but their nature requires further study. Irrespective of the origin of UFOs, we will learn something definitely new and exciting from studying them scientifically. Some of the UFOs are likely real objects whose nature cannot be assessed with any confidence. If they represent alien technology, UFOs could potentially be robotic-autonomous equipment that follows a blueprint crafted by an intelligent species beyond earth. Aliens created equipment that represents our future and would appear magical to us now. The hardware including its equipment visits us, the encounter would echo the actions of higher power. UFO experts say that aliens may or may not be out there, but their technological artifacts exist way beyond our comprehension. The scientific community needs to take a closer look at UFOs by collecting and analyzing new data. In the scientific world the material evidence is important. Could the greatest invention of 21st century be the products of reverse engineering alien technology? Could aliens become the inevitable source of all technology, new materials, and modern invention?

CHAPTER SIX

How did Human Language Evolve?

The evolution of human language has been a controversial topic of debate among scholars, biologists, and anthropologists. It is an intriguing question to which we may never have a complete answer. It is uncontroversial that language has evolved just like any other trait of living organisms. Understanding the evolution of language requires evidence regarding origins and processes that led to change. The most fundamental questions about the origins and evolution of our linguistic capacity remain as mysterious as ever. Language is a gift to mankind by God. However, the origin of language will remain forever obscure. The evolution of human language is an extremely vague topic composed of many different theories and hypotheses with very few instances of evidence to support them. It is generally accepted that evolution of language must have been a long process. All attempts to shed light on the evolution of human language have failed due to the lack of knowledge regarding the origin of language. Language plays a pivotal role in the evolution of human culture. The emergence of language was a defining moment in the evolution of modern humans.

Scholars continue to ponder when and how language emerged as they have for centuries. Language remains in the minds of many philosophers, linguists, and biologists a quintessentially human trait. Our materialistic science is insufficient to explain not only speech came about, but also why we have so many different languages. Despite huge advancements in technology and in our understanding of evolutionary processes since Darwin's theory of natural selection, relatively little is known about the evolution of language. Language evolution shares many features with biological evolution. No one knows exactly how spoken and written language came about or originated. The evolution of language has a lot in common with evolution in organisms. Darwin observed that the formation of different languages and of distinct species, and the proofs that both have been developed through a gradual process. Language is a social institution, which carries culture through art and literature. Linguistic scholars seek to determine what is unique and universal about the language we use, how it is acquired and the ways it changes over time.

Our historians believe that language began 1,50,000 years ago, while written language appeared 6000 years ago. Language has its own words, symbols, and sounds. Language is a communication system comprising sounds, words, and grammar. Language is a system of symbols with several levels of organization, at least phonetics or the sounds, syntax or the grammar, and semantics or the meaning. Language is made up of words and put words together in combination. No one knows for sure when language evolved but genetic data suggests that humanity can probably trace its ancestors back to populations of modern homo sapiens. There are many species of life that are more than able to communicate, but humans have a unique form of communication. We are the only species capable of using language as a form of communication. Human language is distinct from all other known animal forms of communication. Language has played a prominent and possibly pre-eminent role in our species' history. In fact, language is an extension of culture itself and therefore is completely depends on it. However, there is a strong biological element in it.

There are about 7000 languages spoken around the world, and most of us cannot communicate with most other members of our species. Language is a primarily human and non- instinctive method of communicating ideas, emotions, and desires by means of a system of voluntarily produced symbols. In the spoken form, language has been observed to be a product of natural selection, almost Darwinian in nature. Human language is used to exchange specific information with each other about people or objects, and their locations or actions. Language plays an important role in our lives and is very important for human communication. Language is a gift to mankind to survive and evolve. In fact, the evolution of human language is an extremely vague topic, composed of many different theories and hypotheses. Language allows us to share our thoughts, ideas, emotions, and intentions with others. In human language, arbitrary sounds and signs represent specific words, which can be learned, invented, and infinitely combined within the structures of grammar. Language in all aspects, consists of abstract units of information that are organized and combined following specific computational procedures.

We acquire language by living in the society and our family. The conventionalists hypothesize that the relationship between the form of language and meaning was essentially arbitrary, a convention of society. The naturalists hypothesize that the form of a word with sound has a natural association with its referent in the natural world. The survival and preservation of certain favored words in the struggle for existence in natural selection. From never having a single word to over hundreds of thousands of words in modern times, all the words that were created in order for people to communicate are part of a complex system. It is a difficult concept to understand because we have been using language for most of our lives, and all of these words have been in place for thousands of years. Many scholars agree that language began by the use of gestures in primates and hominins. Understanding how the capacity for language evolved is difficult problem, because it is unique to humankind. Languages are not set in stone as they constantly mutate, adding new words and modifying old meanings. The theories of language evolution have been a controversial topic.

Philosophy of language investigates the nature of human language, its origins and use, the relationship between meaning and truth, and how language relates to human thought and understanding, as well as to reality itself. One widely accepted theory is that language came about as an evolutionary adaptation. The evolution of species is a natural process, and so is the language evolution. Most of the languages we have today evolved from the same base tongues spoken thousands of years ago. For the most part, language is passed down through generations so that parents and children are able to communicate through languages, and this extraordinary communication system found exclusively in humans. The evolution of language continues as new generations integrate new technology as a part of their daily , and made thoroughly different entities.

We simply do not know how language originated as well as that spoken language developed well before the written language. All human languages recognize the past, present, and future, and structure words into sentences.

Both vocalization and gesture are universal modes of communication and fundamental feature of language development. Over thousands of years humans have developed a wide variety of systems to assign specific meaning to sounds, forming words and systems of grammar to create languages. The vocabulary and grammar change too rapidly to be able to link all of the world's languages to a common original mother tongue. Humans have a high fidelity code for transmitting information down the generations. Many languages developed written forms using symbols to visually record their meaning. Some languages are incredibly old and have changed very little. The process of word formation enables a language to encode an essentially unlimited number of objects. Language was an innovation that changed radically the character of human society. The process of word formation enables a language to encode an essentially unlimited number of objects. Our mastery of language presents many mysteries, not least, where grammar comes from, and how children learn to speak so effortlessly. By age four, most humans have developed an ability to communicate through oral language.

Numerous theories have been put forward by experts from virtually all disciplines, and yet the problem of how language emerged remains largely unsolved. Language develops according to the needs of culture and civilization, and disappears when that society disappears. Apart from reflecting the reality, language also helps man to perceive and understand the world. Humans acquire language through an instinctive knowledge. Man is called a social animal because of language. Through human interactions new words and phrases are picked up, and integrated into human speech. Language creates a questioning mind, and develops creative thinking. Word formation also takes place by combining informational units in a stepwise process familiar to computational systems. Thought without symbol, and life without language are cognitive reality that are virtually impossible for most modern humans to fathom. There is an another question whether or not a form of language is necessary for thought- do people think in language, or do they think and then translate those thoughts into language. The vale system of a society is all reflected in its language.

There are four theories that explain most of speech and language development: 1. Behavioral theory states that language is a set of verbal behavior learned through operant conditioning. Behaviorists believe that language behaviors are learned by imitation, reinforcement, and copying adult language behavior. 2. Nativists theory is a biologically-based theory, which states that language is innate, physiologically determined, and genetically transmitted. This means that a newborn baby is pre-wired for language acquisition and a linguistic mechanism is activated by exposure to language. 3. Cognitive theory is a perspective of language development that emphasizes the interrelationship between language learning and cognition. Children demonstrated certain cognitive abilities as a corresponding language behavior emerges. 4. Social-Pragmatic theory considers communication as the best function of language. The process of speech and language development in infants and children is complex and interrelated. There must be an integration of anatomy and physiology of the speech systems. Neurological developments, and interactions encourage children for communication attempts.

Mother tongue is the first language that we learn to speak as a child. The term mother tongue refers to a person's native language- that is a language learned from birth. The general usage of the term mother tongue denotes not only the language one learns from one's mother, but also the speaker's dominant and home language. Mother tongue is valuable due to several reasons: it is vital in framing the thinking and emotions. Father tongue is the language spoken by one's father, when it differs from that spoken by one's mother. A mixed language arises among a bilingual group combining aspects of two or more languages, but not clearly deriving primarily from any single language. In most cases, mixed language speakers are fluent, even in native, speakers of both languages. Since the time immemorial Tribal people have practiced, valued, and shared their tribal languages across generation and geography. Tribal language is specific and special to the place and people who speak it. Tribal languages are a treasure trove of knowledge about a region's flora, fauna, and medicinal plants. Tribal languages are spoken by limited people and there is a need to conserve them.

Language contributes in a large way to forming ideas and storing them for future. Facts about language are handed down from person to person from one generation to another through cultural transmission. We use language every day, when we wake up, talk to our beloved ones, go out, and even when we think. Most of the time, we hardly realize that we are talking, thinking, writing, or typing in the spoken word. Language is a complex system and only humans with complex brain system are capable of developing. Most contend that thought and language are two interrelated criteria. Knowing a language means one can speak, be understood, and understand others who know the language. Theorists belonging to cultural roots, believe that cultural understanding is all there to language, with no biological impact. It has been observed that children, who were isolated from socio-cultural sources of language for their formative years, were unable to learn how to speak ever. There are also strong questions regarding the role of intelligence in linguistic abilities. Instinctively babies babble, almost as if they are trying to speak, and eventually learn to talk.

Two basic theoretical positions emerged as explanation frameworks for languages, that which opted for irregularity and that which insisted that language was essentially regular. There are two important basic principles of language: the streamlining effect of least effort, and the compensatory maintenance of communication or restructuring. The most ancient languages for which we have written texts are often far more intricate and complicated in their grammatical forms than many contemporary languages. People seem to tend to re-adapt existing lexical material rather than create entirely new material. Language is considered as a cultural, social, and psychological phenomenon. Although there is a lot in common among languages, each one is unique, both in its structure and in the way it reflects the culture of the people who speak. The factors that contribute to language diversity in different regions of the globe are varied. Researchers hypothesize that language patterns evolve according to history, cultural differences, and geographic divides like mountains or rivers, but there is actually no clear answer with definite direction. Speaking, writing, and reading are integral to everyday life- language is the tool.

Theories of origin of language are generally based on beliefs and hunches. For centuries there had been so much fruitless speculation over the question of how language began. The early theories are now referred to by the nicknames given to them by language scholars. Some of the theories related to the speculative origin of language include: 1. The Bow-Wow theory, which states that speech is the imitation of barking of dogs and other animals. 2. The Pooh-Pooh theory indicates that speech is derived from the automatic emission of painful feelings. 3. The Ding-Dong theory suggests that speech is developed gradually by the harmonization of sound and sense. 4. The Yo- He-Ho theory tells that sound is produced by regulating the breadth as a result of strong muscular action of mouth. 5. The Gesture theory says that humans used their tongue in the same rhythm with gesture and posture to gradually develop language. 6. The Tara- boom- de- ye theory reveals that speech is the result of expression of joy, making different sounds of joy. 7. The La-la theory emerged from the sounds of inspired playfulness, love, poetic sensibility, and song.

Humans currently speak more than 7000 languages. Linguists are sure that the number of languages will constantly change as humanity and the world change as well. Spoken language has always existed prior to written language. What we talk about is limited by constraints on how we see the world, including basic biases affecting how we conceptualize objects and events. Humans have kept moving, and migration, in whatever form and point in time, they might have taken place, helped spread languages with different origins all over the world. When different groups of people initiated contact with each other, they developed languages that were often a mix of the two already spoken ones. Languages have an organic quality to them that allows them to keep changing over time and with who use them. Modern theories of language try to understand how language use arises from domain-general, embodied learning, and process mechanisms that take place in richly structured social environment. In today's world, effective use of language helps us in our inter-personal relationships at home and work. Slang is a very informal language, and today's teenagers have come up with several new slang terms.

The main factor that a language becomes popular is due to a power-base, whether economic, or political, or military. Due to massive British Colonial Conquests, no culture is in complete oblivion of the English language, or words. English is a language that should not appear as too alien, or strange to any community. English is not such big of a deal for most people as they can find a certain level of familiarity with the language. English is a very effective language, and this is evident due to the presence of various native and non-native speakers at a global scale. One-fourth of the world is either fluent in the English language, or content with it. English language has a lot of words and synonyms to express something. As such, any word, or its meaning can be expressed with a high level of accuracy. On a global level, English has the most number of speakers, who speak English either as a first or second language. Further, the language enjoys world wide acceptance and usage by every nation of the world. Without a doubt no other language in the world can come close to English in terms of its immense popularity. English has been the international language of business.

Tamil is the oldest language in the world, and is the mother all indo-European languages .Tamil was spoken in South India for more than 5000 years. It is the oldest extant literature amongst other Dravidian languages. Tamil literature has led to its being described as one of the great classical traditions and literatures of the world. Recent archaeological evidence at Adhichanallur and Indian Ocean shows that the language is related to Sumerians. It is also believed to have numerous great cities with great monuments, and the foremost among them were the two first and second cities of Madurai. The Epics Shilappadikaram, and Manimekhalai describe the submerged city of Puhar. Tamil is the official language of the state of Tamil Nadu, and also has official status in Sri Lanka and Singapore. Tholkappiam is the earliest extant Tamil grammar and literary work, as some of its archaic structures and considerations of style place it earlier than what hs come to be called Sangam Literature. Rishi Agastya is the purported author of the Agasthiyam, a magnum opus and grammar of letters. Thirukkural written by Thiruvalluvar is accepted as a work of great importance written in a masterful style.

Languages have a variety of forms and dialets that can vary in phonology, vocabulary, and grammar within a language. People often ask where words come from or how they come to be. There is no simple or easy way to explain it, except as long as parents and teachers help children to talk. Right from the time children are given birth to, up to the age of five, they develop language at a very fast pace. The mechanisms that enable children to segment syllables and words out of the string of sounds they hear, and to acquire grammar to understand and produce language is still quite an enigma. Words consist of sounds (oral), and shapes (written) that have agreed-upon meanings based in concepts, ideas, and memories. An important aspect to understand about language is that oral language or language used in public speaking, and written language or language used in texts does not function the same way. Oral language has smaller variety of words, shorter sentences, self-reference words, colloquial, and non-standard words, more contractions and more repetition of words and syllables. Written language tends to be more complex and intricate than with speech which is transient.

Society and culture influence the words that we speak, and the words that we speak influence society and culture. We arrive at meaning through conversational interaction, which follows many social norms and rules. Language is central to social interaction in every society, regardless of location and time period. Language shapes social interaction, and social interactions shape language. People adjust the way they talk to their social situation. An individual, for instance, will speak differently to a child than to someone else. How a person chooses to express their own identity has become one of the pivotal battlegrounds of the 21st century. Social media has a tendency to create words of its own at an alarming frequency. The origin of language can be a single proto-language that slowly evolved and was transported to different parts of the world. The origin of language was perhaps the need to communicate. May be the initial words were only howls and hoots, but eventually, they evolved to form a systematic way of communication for humans. Understanding of language evolution will remain one of the greatest mysteries of our species.

Famous Unsolved Mysteries

1. Dark Energy- Dark Energy is a theoretical form of energy postulated to act in opposition to gravity and to occupy the entire universe. Even though gravity is pulling toward on space-time, the fabric of the cosmos keeps expanding outward faster and faster. Dark energy is a hypothetical form of energy that exerts a negative, repulsive pressure behaving like the opposite of gravity. Dark Energy makes up 72 percent of total mass-energy density of the universe. Dark Energy is invisible, fills all of space and its repulsive gravity is speeding up the expansion of the universe. Dark Energy overwhelmed gravity and gained control of the universe about five billion years ago. Astrophysicists have proposed an invisible agent Dark Energy that counteracts gravity by pushing space-time apart. Scientists are puzzled when observations of distant supernovae showed that not only is the universe still expanding, but is accelerating. As space expands, more space is created, and with it more Dark Energy. Dark Energy is really just place holder name for whatever the cause is. Dark Energy is a cosmic constant: an inherent property of space itself. More is unknown than known

about Dark Energy of the universe.

2. Dark Matter- Dark Matter is 27 percent of the universe, which is invisible, and yet the motions of the galaxies suggest it must be there. Most of the matter in the universe is actually unseeable, untouchable, and to this day, undiscovered. Dark Matter as it is called, cannot be seen directly, and has not yet been detected either. The existence of Dark Matter and its properties are inferred from its gravitational effects on visible matter, radiation and the structure of the universe. This shadowy substance is thought to pervade the outskirts of galaxies, and may be composed of weakly interacting massive particles. Dark Matter gives out no light, too little light for us to detect. We know it exists because we see the effects of its gravity on the visible stars, and galaxies. Scientists believe that most of the universe is made up of Dark Matter, which does not emit light or energy. Scientists have some ideas about what Dark Matter might be exotic and still hypothetical particles, but they have hardly any clue about it. Until scientists can make sense of it, the fate of the universe shall remain forever unknown. We are much more certain what Dark Matter is not, than we are what it is.

3. Black Hole- Black Hole is a region of space-time, where gravity is so strong that nothing, including light or other electromagnetic waves, has enough energy to escape it. When a massive star dies, it leaves behind a small dense remnant core, with the force of gravity overwhelming all other forces and produces a Black Hole. Scientists can not directly observe them with telescopes that detect stars. A Black Hole is a great amount of matter packed into a very small area- the result is a gravitational field so strong that not even light can escape. The idea of an object in space so massive and dense that light could not escape has been

around for centuries. Most Black Holes form from the remnants of a large star that dies in a supernova explosion. Black Holes are points in space that are so dense they create deep gravity sinks. Black Holes are some of the strangest and most fascinating objects in space. Anything that ventures too close- be it a star, planet, or spacecraft will be stretched and compressed like putty in a theoretical process. These gluttonous beasts are some of the most fascinating objects in space. Black Holes seem to be the stuff of science fiction, and it is not uncommon for people to wonder.

4. Consciousness- Consciousness has been extraordinarily controversial in the scientific and philosophical traditions. Great thinkers have spent an immense amount of time to unravel mysteries, such as how consciousness works, and where it resides. Many experts agree, however, that conscious beings are aware of their surroundings, themselves, and their own perceptions. Consciousness is also subjective and difficult to define. We do not know how our conscious experiences arise. Consciousness is our awareness of ourselves and the world around us. Science as we know it cannot explain consciousness- one reason is that consciousness is unobservable. Explaining how something as complex as consciousness can emerge from a grey, jelly-like lump of tissue in the head is arguably the greatest scientific challenge of our time. Consciousness is a concept of human experience that is difficult to define because it is not tangible like the human brain. Consciousness is lived reality, and it is the feeling of life itself. What we ultimately want is to explain why conscious experiences are correlated with brain activity. The problem of consciousness, however, is radically unlike any other

scientific problem.

5. Life- We do not have a very good definition for life. It is very abstract thing, what we call life. Some have proposed that life is that which can reproduce. Biologists who devote themselves to the study of living systems remain mystified by what life is. Physicists with their deep understanding of nature's most fundamental laws are less confused. For centuries, scientists and philosophers have proposed hundreds of definitions of life- none have been widely accepted. Astrobiologists are convinced that one cannot successfully search for truly novel forms of microbial life without a definition of life. Life is defined sometimes as cells that self-replicate, metabolize, and are open for mutations. Life is a characteristic of a living organism that distinguishes from a non-living thing. Life is not a problem to be solved, but a mystery to be lived. Some events in life seem meaningful, but looking at our lives as a whole, we realize they do not add up, and it is meaningless. In searching for life in extraterrestrial space, it is essential to act based on an unequivocal definition of life. Evolution is crucial for life to persist on the ever-changing earth. Life is a great and wondrous mystery.

6. Sleep- Sleep accounts for one-quarter to one-third of the human life span. Sleep is a normal, indeed essential part of our lives. Our body craves sleep, much like it hungers for food. Throughout the day our desire for sleep builds, and when it reaches a certain point, we need to sleep. Our body cannot force us eat when we are hungry, but when we are tired, it can put us to sleep. When we are exhausted, our body is even able to engage in micro-sleep. A healthy amount of sleep is vital for the brain's ability to adapt to input. When we do not get enough sleep, our health risks rise, and sleep also plays a role in metabolism. Most of us

acknowledge at some level that sleep makes us feel better. We feel more alert, more energetic, happier, and better able to function following a good night sleep. Getting enough sleep is essential for helping a person maintain optimal health and well-being. It is obvious why we need to eat, but it is not clear why we need sleep at all. It seems that a lack of sleep alters the way in which the genes in the body's cells behave. Despite of decades of research and many discoveries about other aspects of sleep, the question of why we sleep has been difficult to answer.

7. Bicycle- A bicycle is a vehicle with two wheels, which we sit on and ride by moving our legs. Bike is a two-wheeled steerable machine that is pedaled by the rider's feet. It is a vehicle with two wheels tandem, handlebars for steering, a saddle seat, and pedals by which it is propelled. Almost everyone can ride a bicycle, yet apparently no one knows how they do it. As well as fun, cheap, and clean, bikes are an excellent way to witness science in action. Why does a bicycle stay upright while moving? What we may think that the human rider is what keeps a bicycle balanced- this is not completely true. If launched properly, a modern bike with no rider balances itself, just fine and continues on its path as if a rider was guiding it. A bicycle is self-balancing when in motion because of gyroscopic effects and caster effects, alone is the thought of scientists. If we push an unmanned bike forward, it will roll straight without falling over for a little while, at least. Most people do not realize that bicycles can balance themselves. The bicycle stays upright in motion because the wheels keep it balanced. A bike's stability is not easy as it looks. It is a miracle that bicycles can balance at all. The bicyce is still a scientific mystery.

8. Bermuda Triangle- Bermuda Triangle is a mythical section of the Atlantic Ocean roughly bounded by Miami, Bermuda, and Puerto Rico. It is an urban legend focused on a loosely defined region in the Western part of the North Atlantic Ocean, where a number of aircraft and ships are said to have disappeared under mysterious circumstances. Unexplained circumstances surround some of these accidents, including one in which the pilots of a squadron of US Navy bombers became disoriented while flying over the area: the planes were never found. Some speculate that unknown and mysterious forces account for the unexplained disappearances, such as extraterrestrials capturing humans for study, vortices that suck objects into other dimensions, and other whimsical ideas. There is some evidence to suggest that Bermuda Triangle is a place where a magnetic compass sometimes points towards true north as opposed to magnetic north. The ocean has been a mysterious place to humans, and when foul weather or poor navigation is involved, it can be a very deadly place. Mystery around the notorious Bermuda Triangle continues to intrigue scientists as well as regular public.

9. Nazca Lines- Nazca lines are a collection of giant geo-glyphs: designs or motifs etched into the ground depicting various plants, animals, and shapes. They are groups of geo-glyphs, large line drawings that appear from a distance, to be etched into earth's surface on the arid Pampa Colorado in southern Peru. The 2000 year-old Nazca Lines can only be fully appreciated, when viewed from the air given their massive size. There are three types of Nazca Lines: straight lines, geometric designs, and pictorial representations. The Nazca Lines drawn in geometric patterns and distinct animal shapes across the Peruvian desert have inspired many theories over the years. The outstanding geometric

precision of Nazca Lines, transform the vast land into a highly symbolic, ritual, and social cultural landscape that remains until today. These are believed to be the greatest known archaeological enigma, owing to their size, continuity, nature, and quality. One of the most seductive and popular theories involve ancient aliens, who communicated with the Peruvians. The mystery surrounds their exact purpose and the fact that they can only be fully seen from the sky. These Nazca Lines are subject of mystery even for today.

10. Crop Circles- Crop Circle is a geometric and especially a circular pattern of flattened stalks in a field of grain now usually attributed to natural phenomena, or to the work of extraterrestrial beings. Serious studies of Crop Circles have long been hampered by conspiracy theories and the secretive nature of Crop Circle makers-plus scientists' reluctance to engage with a fringe topic. Intricate patterns carved in fields across England in the years 1980s were a vital phenomenon. Also we do not know what factors influenced the location of any Crop Circle. It is only by looking at the field from high above that the real picture is revealed: the whirls and sharp angles that have been pressed into the wheat farm. A growing underground Crop Circle art combines mathematics, technology, stalks, and whimsy. They prompt speculation about alien visitors, ancient spiritual forces, weather anomalies, and other theories. These patterns would appear mysteriously out of nowhere, over night. There has been little in the way of mainstream scientific research on the origins of Crop Circles. Crop Circles have been, perhaps, the best-known geo-spatial mystery of the modern era.

11. Ball Lightning- Ball Lightning is a rare and unexplained phenomenon described as luminescent,

spherical objects that vary from small to several meters in diameter. For millennia people have been telling stories about mysterious spheres of light that glow, crackle, and hover early during thunderstorms. They have been spotted in rural areas, in cities, and on airplanes. They seem out of this world, but scientists believe they are very much of this world. These apparitions are called Ball Lightning, and they remain one of the most mysterious weather phenomena of earth. Ball Lightning usually only last for a few moments, and it is imposible to predict where and when it will show up. It usually occurs near the ground during thunderstorms, in close association with cloud-to-ground lightning. We cannot hunt for Ball Lightning, but it is really Ball Lightning that finds us. It may be red, orange, yellow, white, or blue in colour, and is often accompanied by a hissing sound and distinct odour. It normally lasts only a few seconds, usually moving about, and then vanishing suddenly either silently or explosively. Ball Lightning is one of the strangest phenomena, and unexplained form of lightning on our planet.

12. Moai Statues- Moai Statues are massive megaliths at Easter Island, and these are what this island is famous for. Easter Island is planted with monstrous great statues, the work of whom, we do not know, but its great remains are an enigma. It has nearly 1000 statues, some almost 30 feet tall, and weighing as much as 80 tons. They are still an enigma with the statue builders are far from vanished. Most of the Moai Statues were carted from the volcano Rano Raraku. They were built probably, to honor chieftains or other important people who have passed away, or they were figures of spiritual devotion for the Rapa Nui embodying the spirit of prominent ancestors. They were placed on rectangular stone platforms called ahu, which

are tombs for the people that the statues represent. One of the biggest Easter Island mysteries is how stone- age tribes could succeed in transporting 50 plus ton of Moai Statues kilometers across terrain. These statues have their backs to the sea, keeping watch over the island. Easter Island is the world's most secluded inhabited island, or one of the most remote inhabited locations on earth. It is a quite common belief that aliens made the Moai Statues.

13. Stone Spheres- The mysterious Stone Spheres are found in the deep forests of Costa Rica. They are varied in size from a few inches to seven feet across in diameter, and weighing about 16 tons. There are a lot of myths surrounding the strange Stone Spheres. While some believe the stones were brought to earth by aliens, others are of the view that they have a connection with the lost continent of Atlanta. The Stone Spheres are distinctive for their perfection, their number, size, and density. There are nearly 300 giant Stone Spheres perfectly-formed of granite-like igneous rock. Each of the Stone Spheres appears to be almost perfectly round and is very smooth. The most interesting thing about the Stone Spheres is the precision with which the rocks are given the round shape. We really do not know why they were made and for what purpose. The exact purpose of the Stone Spheres remains a mystery to this day. Their origin, exact age, and history are still a mystery. Scientists are not sure about the use of these Stone Spheres. The Stone Spheres of Costa Rica are one of the strangest mysteries in archaeology. Where in the world did these Stone Spheres come from?

14. Stonehenge- Stonehenge is arguably one of the most megalithic monuments in the world. Stonehenge is a massive stone monument located on a chalky plain north of modern- day city of Salisbury, England. Researchers show

the structure that we call Stonehenge was built nearly 5000 years ago. The biggest Stonehenge's stones known as sarsens are up to 30 feet tall and weigh 25 tons. It is widely believed that they were brought from a distance of 32 kilometers. Smaller stones referred to as bluestones weigh up to 4 tons. It is unknown how people of antiquity moved them that far. Many theories have been put forward so to why Stonehenge was constructed. It is also one of the most mysterious with its prehistoric concentric rings garnering plenty of speculation as to why and how they were constructed. Some argue that Stonehenge is a spacecraft landing area for aliens. Other theories put forward to the plausible use of Stonehenge seem to be a sacred burial site, a site for celestial or astronomical alignment, or a place of healing. Stonehenge itself is the most mysterious with its prehistoric concentric rings gathering plenty of speculation.

15. Pumapunku Monuments- Pumapunku is the name of a temple complex located near Tiwanaku, in Bolivia. The temple's origin is a mystery, and is a part of a larger archaeological site known as Tiahuanacu built around 5000 years ago. The most intriguing thing about Pumapunku is the stonework: the red sandstone are cut in such a precise way that they fit perfect into and lock with each other without using mortar. The technical finesse and precision displayed in these stone blocks is astounding. Not even a razor blade can slide between the rocks. Some of these blocks are finished to machine quality and the holes drilled to perfection. Each stone was finely cut to interlock with the surrounding stones and the blocks fit together like a puzzle, forming load bearing joints without the use of mortar. The stones are of mammoth proportion, the largest of these blocks is 25.6 feet long, 17 feet wide, and 3.5

feet thick and is estimated to weigh 131 metric tons. It is indicative of highly sophisticated knowledge of stone-cutting and geometry. Enduring mystery surrounds the ancient monument s of Pumapunku. Ancient alien theorists and parapsychologists have hypothesized alien intervention.

16. Machu Picchu- Machu Picchu is a 15th century Incan citadel set high in the Andes mountains in Peru. It stands 2430 meters above sea-level in the tropical mountain forest, in an extraordinarily beautiful setting. This historic sanctuary of Machu Picchu is among greatest artistic, architectural, and land use achievements anywhere, and the most significant tangible legacy of the Inca civilization. Nearly, more than 200 structures making up this outstanding religious, ceremonial, astronomical, and agricultural centre is set on a steep ridge, crisscrossed by stone terraces. It is mortar- free limestone architecture. The structures were so well built with a technique called ashlar: the stones that were cut to fit together without mortar. Machu Picchu has a number of structures that would have enhanced the spiritual significance of the site. One of them, the temple of sun has an elliptical design and is located near where the Inca emperor is believed to have resided at Machu Picchu. There is an indication of alignment between the temple rock and solstice sun. Its origin is still the subject of study, and to this day many of Machu Picchu's mysteries remain unresolved.

17. Antikythera Mechanism- Antikythera is an ancient celestial calculator, and the device tracks the motion of the sun, moon, and five planets. It was discovered by sponge divers off the Greek island of Antikythera in 1901, the remains of the mechanism are now preserved in the National Archaeological Museum in Athens. The recovered

fragments of the mechanism contained writing and inscriptions. It is sometimes called the world's oldest computer for its ability to perform astronomical calculations. It was a mechanical computer of bronze gears that used ground-breaking technology to make astronomical predictions. The mechanism represents a level of technology exceeding anything else of the kind for which we have either physical remains or detailed descriptions from antiquity. It is strongly suggestive of an ancient Greek tradition of complex mechanical technology. Scholars have been trying to understand the device but many questions remain unanswered. Antikythera Mechanism is one of the world's oldest known geared devices that has puzzled and intrigued historians of science and technology. This remarkable device seems likely to remain forever a mystery.

18. Atlantis City- the Lost City of Atlantis is one of the oldest and greatest mysteries of the world. Since ancient times, people have been trying to locate Atlantis, which is believed to have submerged after an earthquake or tsunami. Greek philosopher Plato described Atlantis as a large island located near the Rock of Gibralter, home of the most advanced civilization, and being of unrivaled refinement with glorious palace. Atlantis was filled with beautiful citizens, a Poseiden temple, concentric walls, and canals. Even after years of research, the exact truth about this city has not been found and that adds even more to all the folklore attached to it. He fertile and beautiful City of Atlantis, where half-god, and half-human lived, is believed to have been an autonomous region, where people grew their food and reared animals. It is also speculated that the original inhabitants of the Lost City of Atlantis are believed to be of extraterrestrial origin. The inhabitants are believed

to possess exceptional powers such as the ability to control the weather and modify volcanic eruptions. The myth of Atlantis has demonstrated a remarkable persistence over the millennia.

19. San Luis Valley- San Luis valley is a region in south-central Colorado. A supernatural and mysterious, the San Luis Valley is one of the more unusual and beautiful parts of Colorado. When we set foot in the valley, we get a sense that there is more than meets the eye. With its mysterious lure and spirit, it is no wonder that the valley is home to a unique vortex of energy. This largest alphine valley in the world has been inhabited by aliens for thousands of years, and UFO sightings are a common occurrence here. A flying humanoid was spotted in the desert in 2009. There might be something extraterrestrial going on in Colorado' San Luis Valley-at least what some of the long-term residents tell us. There is a UFO watchtower- a humble deck-like platform made of metal 10 feet above the ground. There are reports of dots of light moving at impossible speeds in erratic direction. The snaking stone of unknown origin measuring hundreds of feet long is the weirdest feature of this mysterious valley. The bizarre fact is that the northern part of the valley floor contains Great Sand Dunes with the sand contains some of the purest silica with a distinct mineral composition on earth.

20. Skinwalker Ranch- Several of the researchers in Skinwalker Ranch of northeastern Utah, claim to have seen paranormal activities with incredible stories. The place is filled with extraordinary beliefs of mysterious sightings and encounters. Some have called it a supernatural place with reports persisting of UFOs, crop circles, cattle mutilation, and shape shifting creatures. The crop circles were designed in a triangular pattern. There is no denying

the bizarre extraterrestrials or paranormal activity rich with conspiracy. There have been many reported sightings of UFOs, big foot-type creatures, animal mutation, and unexplained lights. According to some observers, odd objects have been spotted overhead, while strange fireballs have been spotted over the sky. Skinwalker Ranch and the surrounding area have been referred to as UFO alley since 1950s, when numerous anomalous events and strange activity began occurring. The area of Skinwalker Ranch has become a paranormal hotbed of UFO research. It is a living laboratory of so much unexplained unique experiences. It is the birth place of monster tales, alien spaceships, and other strange events.

21. Stone Jars- The unusual site of thousands of huge megalithic ancient Stone Jars scattered across various places in the mountains of Laos has fascinated archaeologists as well as scientists. Ranging in size from three to ten feet and weighing up to 14 tons a piece, they have attracted many scholarsLooking into make sense of their origin story. A nearby cave with human remains, suggests the jars may have been used as funeral urns for chieftains. Another theory suggests that they could have been used as ancient water bottles, possibly for rainwater collection for travelers to use during dry season. The Stone Jars are believed to be around 3000 years old, and most of them were sculpted from sandstone rock. Believed to have once stored either human remains or rice wine, thousands of mysterious Stone Jars lie in ruin. Human skeleton remains were found inside and also buried around the jars. Local legend tells that they were made by a race of giants to use as cups to drink rice wine. The stone source is not found in the immediate vicinity of the site. These jars have been studied for decades, yet mysterious details continue to

puzzle researchers.

Author Bio

Prof R V M Chokkalingam

Prof. RVM. Chokkalingam is a former lecturer/ curator/ scientist, and now @ 80, a retired local professor living in Bangalore. He is a science museum scholar, specialized in the design of Science exhibition@ Science Museum, London. He has made lifetime contribution towards Public Awareness of Science for more than 55 years now. His special contribution towards Public Outreach Programs include: mobile science exhibition, low-cost teaching aids in physics, science Fair Projects, and establishment of NAL museum. He has published more than 170 articles in newspapers and magazines, and 23 books on science, philosophy, nature and digital world. He is the recipient of Karnataka State Award for Science Communication in 2012. This book has arisen as a consequence of author's obsession with the unexplainable mysteries of science.